Pierluca Coiro

Plasticity-related gene 5 induces spines in immature primary neurons

AF003324

Pierluca Coiro

Plasticity-related gene 5 induces spines in immature primary neurons

A role of PRG-5 in dendritic spine formation

Südwestdeutscher Verlag für Hochschulschriften

Impressum / Imprint

Bibliografische Information der Deutschen Nationalbibliothek: Die Deutsche Nationalbibliothek verzeichnet diese Publikation in der Deutschen Nationalbibliografie; detaillierte bibliografische Daten sind im Internet über http://dnb.d-nb.de abrufbar.

Alle in diesem Buch genannten Marken und Produktnamen unterliegen warenzeichen-, marken- oder patentrechtlichem Schutz bzw. sind Warenzeichen oder eingetragene Warenzeichen der jeweiligen Inhaber. Die Wiedergabe von Marken, Produktnamen, Gebrauchsnamen, Handelsnamen, Warenbezeichnungen u.s.w. in diesem Werk berechtigt auch ohne besondere Kennzeichnung nicht zu der Annahme, dass solche Namen im Sinne der Warenzeichen- und Markenschutzgesetzgebung als frei zu betrachten wären und daher von jedermann benutzt werden dürften.

Bibliographic information published by the Deutsche Nationalbibliothek: The Deutsche Nationalbibliothek lists this publication in the Deutsche Nationalbibliografie; detailed bibliographic data are available in the Internet at http://dnb.d-nb.de.

Any brand names and product names mentioned in this book are subject to trademark, brand or patent protection and are trademarks or registered trademarks of their respective holders. The use of brand names, product names, common names, trade names, product descriptions etc. even without a particular marking in this works is in no way to be construed to mean that such names may be regarded as unrestricted in respect of trademark and brand protection legislation and could thus be used by anyone.

Coverbild / Cover image: www.ingimage.com

Verlag / Publisher:
Südwestdeutscher Verlag für Hochschulschriften
ist ein Imprint der / is a trademark of
AV Akademikerverlag GmbH & Co. KG
Heinrich-Böcking-Str. 6-8, 66121 Saarbrücken, Deutschland / Germany
Email: info@svh-verlag.de

Herstellung: siehe letzte Seite /
Printed at: see last page
ISBN: 978-3-8381-3194-8

Copyright © 2012 AV Akademikerverlag GmbH & Co. KG
Alle Rechte vorbehalten. / All rights reserved. Saarbrücken 2012

"Back in the 1960s and early '70s I took plenty of LSD. A lot of people were doing that in Berkeley back then. And I found it to be a mind-opening experience. It was certainly much more important than any courses I ever took."

Kerry Mullis

Acknowledgement

I would like to express my deepest and sincere gratitude to Frau Juniorprof. Anja Bräuer for her supervision, guidance and the trust shown me in recent years.

I would to thank Dr. Tania Velmans for the results on figures 4.10 – 4.11 and 4-12.

I would to thank all the people in my laboratory: Rike Dannenberg, Nora Ebermann, Jan Csupor, Miam Petzold and Anna Soriguera.

I'd like to dedicate my manuscript to Riccardo, Anette, Alessandro and Rakel.

Index

Index	V
Abbreviations	VII
Summary	IX
1. Introduction	1
1.1 Plasticity-related genes	3
1.2 Plasticity-related gene 1	4
1.2.1 Up stream region of PRG-1: a "proximal promoter"	6
1.3 Plasticity-related gene 5	7
1.3.1 Spine formation and hypothetical role of mouse PRG-5	8
2. The aim of work	14
3. Material and methods	15
3.1 Material	15
3.1.1 Laboratory equipment	15
3.1.2 Solution and buffer	16
3.1.3 Biological substances and Kits	18
3.1.4 Software	19
3.2 Methods	20
3.2.1 Vector construction of PRG-1	20
3.2.2 Restriction enzymes	20
3.2.3 Agarose gel electrophoresis	20
3.2.4 Purifying linear DNA	21
3.2.5 Isolation of plasmids: mini prep	21
3.2.6 Bacterial cells	21
3.2.7 Transient transfection & duel luciferase assay	22
3.2.8 Statistical analysis	22
3.2.9 RNA isolation	22
3.2.10 DNAse treatment of total RNA sample and RNA cleanup	23
3.2.11 cDNA synthesis	23
3.2.12 PCR	23
3.2.13 Molecular cloning of PRG-5 and sequence analysis	24
3.2.14 Ligation and transformation	26

	3.2.15 Clone analysis	27
	3.2.16 Site direct mutant of mouse PRG-5	27
	3.2.17 HEK-293, preparation and transfection	29
	3.2.18 Primary neurons, preparation and transfection	29
	3.2.19 qRT-PCR	30
	3.2.20 Immunocytochemistry	31
	3.2.21 In situ hybridization	31
	3.2.22 SDS-polyacrylamide-gel electrophoresis (Protein)	32
	3.2.23 Western blot	32

4. Results — 33
4.1 Identification of proximal promoter of PRG-1 — 33
4.2 Expression pattern of PRG-5 — 36
4.3 Sub-cellular localization of PRG-5 — 42
4.4 PRG-5 induces spine formation in primary neurons — 44
4.5 Analysis of residues within C1 – C3 domains of PRG-5 — 49

5. Discussion — 55
5.1 Proximal promoter of PRG-1 — 55
5.2 Plasticity-related gene 5 (PRG-5) — 57

6. Bibliography — 63

Abbreviations

aa	Amino acids
AMPA	α-amino-3-hydroxyl-5-methyl-4-isoxazole- propionate
ATX	Autotaxin
bHLH	Basic helix-loop-helix
CA-1	Cornus ammonis-1
C1 – C3	Consensus phosphatase sequence
C1P	Ceramid-1-phosphate
CNS	Central nervous system
DIV	Days in vitro
dNTPs	Deoxynucleotide Triphosphate
DPE	Downstream promoter element
DTT	Dithiothreitol
E	Embryonal day
EGFP	Green fluorescence protein
GAPDH	Glyceraldehyde-3-phosphate dehydrogenase
GFAP	Glial fibrillary acid protein
Glu	Glutamate
GTFs	General Transcription Factors
h	Human
HPRT	Hypoxanthine-guanine phosphoribosyltransferase
Inr	Initiator sequence
IB	Immunoblot
Ipa1	Importin alpha 1
LPA	Lysophosphatidic acid
LPC	Lysophosphatidylcholine
LPP	Lipid phosphate phosphatase
LPRs	Lipidphosphatase-related proteins
LTP	Long-term Potentiation

m	mouse
MAG	Monoacylglycerol
NMDA	N-methyl-D-aspartic acid
P	post-natal day
PA	Phospholipase A_2
PAP2D	Phosphatidic acid phosphatase type 2D
PBS	Phosphate buffer
PCR	Polymerase chain reaction
PFA	Paraformaldehyde
PIC	Preinitiation complex
PLD	Lysophospholipase D
PRGs	Plasticity-related genes
r	Rat
RLM	RNA ligase-mediated
S1P	Sphingosin-1-phosphate
TBP	Transcription Binding Protein
TFII	Transcription Factor II
TSS	Transcription start site
YPF	Yellow Fluorescent Protein
WT	Wild type

Summary

The PRG family, a set of five trans-membrane proteins, were shown to be vertebrate specific and mainly expressed in brain tissue. Their actions include neurite extension, axonal path finding and reorganization after lesion. In particular, deletion of PRG-1 results in severe hippocampal overexcitability, while overexpression of PRG-3 in HeLa and COS7 cells increases the physiological number of filopodia. In the central nervous system, PRG-1 expression is restricted to postsynaptic dendrites on glutamatergic neurons.

In this study, first of all I shortly focus my attention in the upstream region of mouse and human PRG-1; PRG-1 expression is under the control of a TATA-less promoter with multiple transcription start sites. The promoter architecture of PRG-1 resulted in the identification of
450 bp, mediating approximately 40-fold enhancement of transcription in cultured primary rat neurons, compared to controls.

Afterwards, I characterized the last member of PRG family, plasticity-related gene-5 (PRG-5), which is able to induce spine-like structure in young primary neurons.

The shape of a neuron's dendritic arbor determines the set of axons with which it may form synaptic contacts, thus establishing connectivity within neural circuits. Dynamic cytoskeleton remodeling is an essential step during this process. Putative extracellular cues may act through membrane proteins that relay signals to a network of intracellular signaling pathways, which ultimately converge on the cytoskeleton. However, the molecular mechanisms involved in these steps are not well understood.
The novel member of the vertebrate- and brain-specific PRG family, plasticity-related gene 5 (PRG-5), a multi-spanning membrane protein, localizes to and promotes the induction of spines in young primary neurons. A set of amino acid within the C1 – C3 domains are responsible for spine formation in primary neurons at DIV 2 and DIV 4.

In fact, mutagenesis experiments in PRG-5 show residual amino acids that are important for the induction of spines. Our data show that PRG-5 may be involved in spine induction in primary neurons and thereby modulate plasma membrane rearrangement.

1. Introducion

The plasticity-related gene family (PRG), comprising integral multi-spanning membrane proteins, is a subclass of the lipid phosphate phosphatases (LPP) superfamily, which controls the level of lipids in the extracellular space (27, 29, 30, 50) and the regulation of effects induced by lysophosphatidic acid (LPA) (6).

LPA is a simple lysophospholipid, a metabolite in the biosynthesis of membrane phospholipids and a well-characterized signaling molecule (37). LPA is present in biological fluids and activates cells through families of G-protein-coupled receptors. This leads to multiple biological functions including cell proliferation, membrane depolarization, and cytoskeletal remodeling in numerous cell types (37). Extracellular LPA is produced in inflammatory conditions by secretory phospholipase A_2 acting on PA (21). LPA is also partially derived through an extracellular lysophospholipase D (lysoPLD) from a major circulating lipid, lysophosphatidylcholine (Fig. 1.1). This lysoPLD has been identified as autotaxin, an ecto-phosphodiesterase (68).

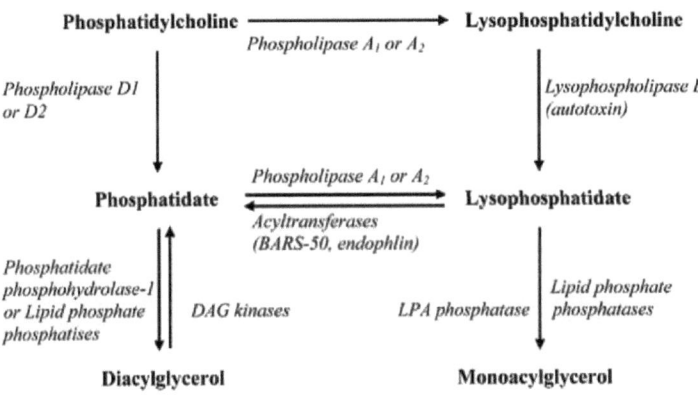

Fig 1.1 Metabolism of phosphatidate and lysophosphatidate. Modified from (47).

Increasing the degradation of extracellular LPA, or attenuating signaling through the LPA receptors can regulate the extracellular actions of LPA providing a control of diverse signaling cascades.

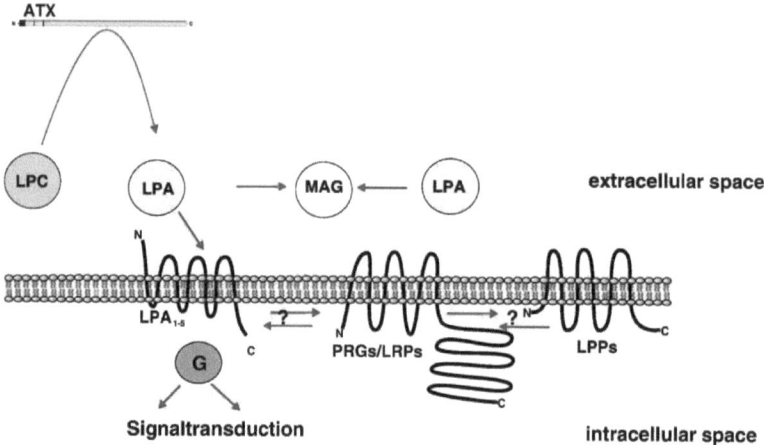

Fig. 1.2 LPA metabolism. A major route for the synthesis of extracellular LPA is through the action of ATX by conversion of lysophosphatidylcholine (LPC) to LPA. The role of LPA in controlling signal transduction is regulated by the ecto-enzymes lipid phosphate phosphatases (LPPs), which are responsible for its degradation to monoacylglycerol (MAG) and also for controlling signaling events downstream of receptor activation (6).

Lipid phosphate phosphatase-1 (LPP1) regulates the degradation of extracellular LPA as well as the intracellular accumulation of lipid phosphates (46). More in general, the effects of the members of the lipid phosphate phosphatases superfamily, LPP1-3 are in the direct opposition to those of ATX. LPPs show a ubiquitous expression pattern (31, 63) and the mRNA are also expressed in the brain, whereas levels of LPP-2 transcripts are somewhat lower and more restricted (27, 31, 63).

All the member of the LPP superfamily are characterized by six membrane-spanning domains with three extracellular loops (27, 29, 30, 50) and are able to dephosphorylate LPA to monoacyl-glycerol (MAG) (Fig. 1.2) (9, 10, 74). The conserved enzymatic active domains are localized within the three extracellular loops. The conserved amino acids, essential for an ecto phosphatase activity, are shown in Tab 1 (74). LPP-1 is partially expressed on the plasma membrane of cells with the C- and N- terminal inside the cell (30) and a glycosylation site on a hydrophilic loop between the first and second active site domains (9, 49, 57). LPP-1, LLP-1a and LPP-3 are partly located in the plasma membrane of some cells (29, 30, 39, 53). Their observed ability of the LPPs to dephosphorylate exogenous lipid phosphates (29, 30) implies that the LPPs could regulate circulating concentrations of LPA or S1P in biological fluids.

1.1 Plasticity-related genes

Recently, a family of plasticity-related genes (PRGs), or LPP-related proteins (LPRs) that show structural similarity to the LPPs has been identified (Fig. 1.3). These proteins, like LPPs, all contain six putative transmembrane regions. Unlike other members of the LPP family, PRG-1 and PRG-2 (LPR-4 and -3, respectively) have very long hydrophilic C-terminal tails.

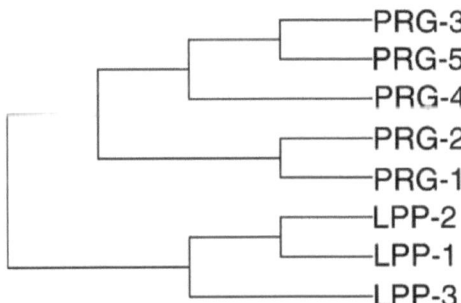

Fig. 1.3 Phylogenetic tree between mouse LPPs and mouse plasticity-related genes (PRGs) calculated using the Clustal W algorithm (6).

Mouse PRG-1 (GeneBank Accession no. AAP41099) protein contains 766 aa, PRG-2 (NP_859009) 746 aa, while PRG-3 (NP_848871), PRG-4 (Q8VCY8) and PRG-5 (AAS80161) are 325 aa, 343 aa and 321 aa, respectively. Although all PRGs protein show a high identity to LPP-family members, included the same six spanning membrane structure, they lack the residues responsible for a phospho enzymatic activity (Tab 1) (26, 61, 64, 74). Nevertheless, PRGs are involved in a variety of neuronal processes like filopodia formation, neurite extension, axonal path finding and reorganization after lesions (7, 43, 51, 58).

		C1		C2		C3
mPRG-1	(206)	FYFLTVCKP	(251)	SQH	(297)	TRITQYKNHPVD
mPRG-2	(158)	PFFLTVCKP	(203)	SQH	(249)	TQITQYRSHPVD
mPRG-3	(154)	FYFLTVCQP	(198)	SKH	(245)	NRVAEYRNHWSD
mPRG-4	(156)	PHFLSVCRP	(207)	CKD	(254)	VRVAEYRNHWSD
mPRG-5	(149)	PHFLALCKP	(193)	SKE	(240)	NRVAEYRNHWSD
mLPP-1	(128)	PHFLAICNP	(169)	SGH	(216)	SRVSDYKHHVSD

Tab. 1 The consensus phosphatase sequence motif is shown in alignment between PRGs and LPP-1 of mouse (58).

In details, through my thesis, I investigated two distinct aspect of a gene regulation: the upstream genomic sequence region of PRG-1, indicated as a putative "proximal promoter" and may be responsible of neuron specific expression of PRG-1.
Then, I characterized the new member of PRG family, named PRG-5. The overexpression of PRG-5 is able to induce spine structure in young primary neurons.

1.2 Plasticity-related gene 1

PRG-1, the first member of the PRG-family, was discovered in a screen to identify proteins involved in reorganization in the dentate gyrus, following entorhinal cortex lesion, which is an established model for the analysis of lesion-induced plasticity (52, 59). The six spanning membrane structure is showed in Fig. 1.4.

The mRNA of PRG-1, prominently expressed in the CNS and vertebrate-specific, was detected as early at embryonic day 19 (E19) in the subventricular zone and hippocampal anlage. After birth, PRG-1 transcripts are strongly up regulated in the hippocampus and entorhinal cortex, and the protein is thus found exclusively on glutamatergic neurons in the CNS (6).

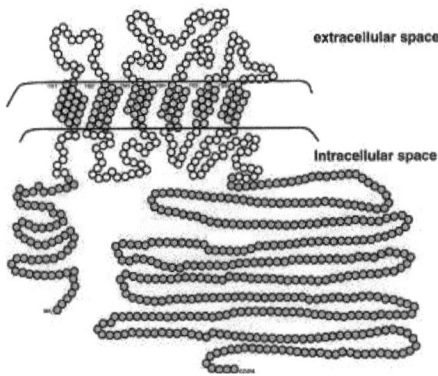

Fig. 1.4 Structure model based on the amino acid sequence of mouse PRG-1. PRG-1 contains six putative transmembrane regions and the C- and the N-terminus located in the intracellular region. Modified from (7).

Analysis of PRG-1 protein during excitatory synaptic transmission is thought to depend on an interaction of PRG-1 and lipid phosphates like LPA, acting through presynaptic LPA2-receptors (66).

1.2.1 Up stream region of PRG-1: a "proximal promoter"

Transcription of eukaryotic protein-coding gene is preceded by multiple events; these include decondensation of the locus, nucleosome remodeling, histone modifications, binding of transcriptional activators and coactivators to enhancers and promoters, and recruitment of the basal transcription machinery to the core promoter. Core promoter elements define the site for assembly of the transcription preinitiation complex (PIC) and include TATA sequence, located upstream of the transcription start site (TSS), and an initiator sequence (Inr), encompassing the start site. Promoters can include a TATA box, an Inr sequence, or both of these control elements. A third core element, the downstream promoter element (DPE), was initially described in *Drosophila* and is located about 30 bp downstream of the start site (12). The DPE appears to function, in conjunction with the Inr element, as a TFIID binding site at TATA-less promoters. The core promoter includes also DNA elements that can extend ~35 bp upstream and/or downstream of the transcription initiation site (TSS). Most core promoter elements appear to interact directly with components of the basal transcription machinery. The basal machinery can be defined as the factors, including RNA polymerase II itself, that are minimally essential for transcription in vitro from an isolated core promoter (24, 41, 70) (Fig. 1.5).

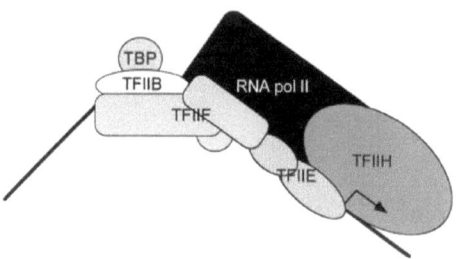

Fig. 1.5 Schematic description of the transcription preinitiation complex (PIC).

PIC assembly is nucleated by TBP binding to the TATA box, inducing a sharp bend in the DNA template, followed by association of TFIIB, RNA pol II/TFIIF, TFIIE, and TFIIH. Each pattern denotes a distinct general transcription factor. Subunit composition is indicated, except for TFIIH (9 subunits) and RNA pol II (12 subunits). Although PIC assembly can occur by stepwise addition of the general transcription factors (GTFs) in vitro, the discovery of RNA pol II holoenzyme complexes that include GTFs suggests that stepwise assembly might not occur in vivo (24).

Although core promoters for RNA polymerase II were originally thought to be invariant, they have been found to possess considerable structural and functional diversity (60). Furthermore, it appears that core promoter diversity makes an important contribution to the combinatorial regulation of gene expression (60).

PRG-1 was found exclusively on glutamatergic neurons in the CNS. Analysis of PRG-1 null mouse models elucidated the modulatory role of the postsynaptically localized PRG-1 protein during excitatory synaptic transmission on glutamatergic neurons. This novel molecular mechanism of adaptation in neuronal transmission is thought to depend on an interaction of PRG-1 and lipid phosphates like LPA, acting through presynaptic LPA_2-receptors (66).

1.3 Plasticity-related gene 5

PRG-5, previously described as phosphatidic acid phosphatase type 2D (PAP2D) (62), was recently found, like PRG-3 (58), to increase the number of filopodia in N1E-115 neuronal cells (11). PRG-5 is located inside the same chromosome of PRG-1 (chromosome 4 in human and chromosome 3 in mouse), consecutively and in an opposite orientation (Fig. 1.6) and it shows the same six spanning membrane structure, with intracellular N- and C- terminus. Overexpression studies will be performed to show the capacity of PRG-5 to induce spine structure in immature primary neurons.

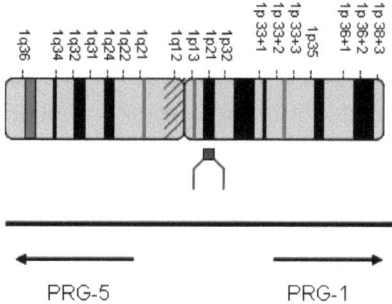

Fig. 1.6 Schematic model of chromosome 3 in mouse, with opposite orientation of PRG-1 and PRG-5.

1.3.1 Spine formation and hypothetical role of mouse PRG-5

Neuronal activity comprises an enormous number of contacts between somata, axons and dendrites, known collectively as synapses. Synapses regulate the electric communication within neural networks and most excitatory synapses in the mammalian brain are formed from tiny dendritic protrusions, named dendritic spines (4) (Fig. 1.7).

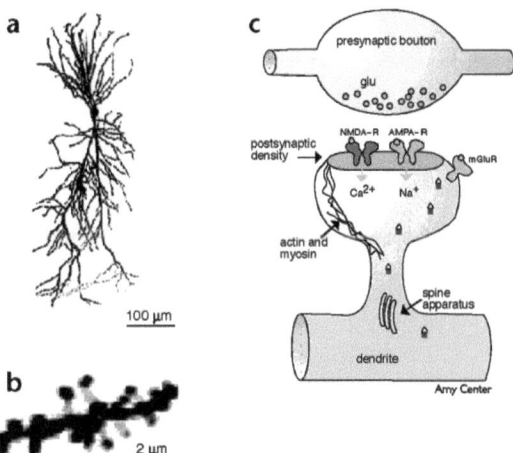

Fig. 1.7 (a) A CA-1 pyramidal cell from a hippocampal slice. The cell is filled with Lucifer Yellow, scanned and reconstructed by a confocal laser-scanning microscope. (b) Example of dendritic segment with spines, selected from the cell in (a). (c) Diagram of the presynaptic bouton and a postsynaptic spine. The spine head carries three classes of glutamate receptors: the metabotropic glutamate receptor (mGluR), the ionotropic AMPA receptor (AMPA-R) and the ionotropic NMDA receptor (NMDA-R). The bouton contains transmitter vesicles with glutamate (glu). The red squares represent proteins carried by a lipid vesicular transport system to support membrane expansion and functional adjustment of spines (1).

Five main morphological stages can be identified in dissociated cultured of hippocampal neurons (16) (Fig. 1.8). Shortly after the cells attach the substrate, motile lamellipodia develop around the periphery of the cell (stage 1). The significance of lamellipodia in neuronal development is unclear (16). At stage 2, development is characterized by the transformation of lamellipodia into distinct processes, which, in the course of a few hours, extend to a length of 10-15 µm (56). Several hours after the appearance of minor processes, a rather abrupt change occurs: one of the minor processes begins to grow at a much more rapid rate. From this stage onward (stage 3), its rate of growth will average 5-10 times greater than the other processes of the cell. This process is the axon. At stage 4, like the axon, dendrites develop from the minor processes that appear during the first day in culture, but significant dendritic growth begins only after about 4 d in culture, 2-3 days after axonal outgrowth (16).

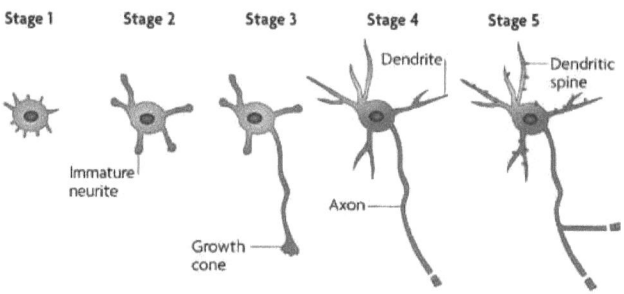

Fig. 1.8 Schematic representation of neuronal polarization in cultured rat embryonic hippocampal neurons. Shortly after plating, the neurons form small protrusion veils and a few spikes (stage 1). These truncated protrusions have growth cones at their tips, and develop into several immature neurites (stage 2). One neurite then starts to break the initial morphological symmetry, growing at a rapid rate, and immediately establishing the polarity (stage 3). A few days later, the remaining neurites elongate and acquire the characteristics of dendrites (stage 4). Approximately seven days after plating, neurons form synaptic contacts through dendritic spines and axon terminals, and establish a neuronal network (stage 5) (16).

Despite different knowledge of dendritic spine were described, key aspects remaining unclear, as "how do spines form?". Several lines of evidence suggest that dendrites produce long and thin dendritic filopodia that exhibit dynamic growth, allowing them to capture some of the nearby axons. (16, 54) (Fig. 1.9). Choosing and capturing the appropriate presynaptic axon via activity-dependent or -independent signaling stabilizes the contact and the filopodia mature into dendritic spines. Filopodia grow and retract within minutes or even seconds and their motility is regulated by synaptic activity (3, 17, 33, 48, 69, 73).

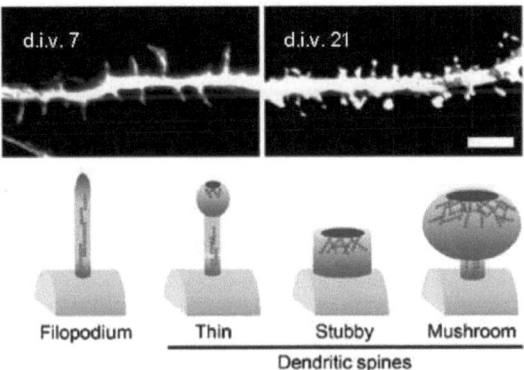

Fig. 1.9 Morphology of dendritic protrusions, filopodia and spines. Dendrites of GFP-transfected hippocampal neurons cultured for 7 days (d.i.v.7) and 21 days (d.i.v. 21). At the immature stage (7 days), dendritic protrusions are very thin and long; these protrusions are called dendritic filopodia. In contrast, at the mature stage (21 days), dendrites are covered by dendritic spines, which commonly have an expanded head and a narrow neck. Scale bar, 5 mm. Schematic representation of morphologies of filopodia and three types of dendritic spine: thin type, stubby type and mushroom type. Gray disks represent the PSD structure and chains of red circles represent F-actin (54).

Furthermore, time-lapse imaging studies have found that spines can form directly from dendritic shafts rather than transitioning from filopodia (15).
Small spines can change their form rapidly (22, 42, 65), either disappearing or growing into large spines. By contrast, large spines are relatively stabled *in vitro* (42) and survive for more than a

month (65) or even for a year (22) in the mouse neocortex *in vivo*. However, these studies do not directly address the correlation between spine structure and function.

The in vivo imaging of YFP-expressing mice showing that dendritic filopodia are indeed highly dynamic and can transform into spines provides more evidence consistent with the role of filopodia in spinogenesis (22, 76). Normally, a mice at one month of age show ~12% of filopodia; in almost 4h, the 15% of filopodia form a bulbous head and more the 40% of these spine-like protrusions persist more than 24h (76). These evidences provide that filopodia are spine-precursors.

Although, that result give credence to that theory, other models of spinogenesis may also occur and should be considered. Shaft synapses can precede the formation of spine synapses, during early developmental stages (19, 73). Furthermore, time-lapse imaging studies have found that spines can form directly from dendritic shafts rather than transitioning from filopodia (15). Thus, it could be possible that presynaptic axons recognize the shafts of dendrites and induce the postsynaptic cell to form a dendritic spine directly in apposition to the axonal terminal (19, 73). Moreover, spines may emerge independently of synaptic contact, acting as beacons for axonal terminals, which would then locate and recognize these preformed spines on dendrites and make targeted contacts. To support these model, it is interesting consider that in some parts of the brain, spinogenesis can occur in the absence of synaptogenesis (8, 13). Even if the prominent idea remain that spines came from a development of dynamic filopodia, it is still unclear if filopodia are always precursor of all spines or if most spines form themselves, directly from the dendritic shaft.

Recently, various studies on the components regulating spine morphology, such as long-term potentiation (LTP), induction of LTP, or motility of spines, have been found to be associated with the remodelling of actin cytoskeleton (20, 23, 25, 36). According to these studies, LTP increases the number of spines in the hippocampus (18), while overexpression of Rac1 induces changes in the morphology of spines (38). Spine-like structures from neurons expressing EGFP alone showed altered morphology after overexpression of Kalirin-7 in primary cortical neurons cultured for DIV 7, exhibiting a high number of protrusions of larger size and more complex shape than control neurons (45). Morphogenesis and maturation of dendritic spines are regulated upstream by the activation of the ephrinB-EphB receptor, which regulates the translocation of Rho-GEF Kalirin and the activation of Rac1 (44). Downregulation of the protein Septin 7, normally expressed in cultured neurons between DIV 0 and DIV 20, affects dendritic-spine morphogenesis (71) and loss of function of WAVE1 (Wiskott-Aldrich syndrome protein (WASP)-family verprolin homologous protein 1) reduces the number of mature dendritic spines (32). WAVE1 also regulates the formation of the filamentous actin cytoskeleton by phosphorylation/dephosphorylation processes. Activation of WASP protein by cdc42 and Arp2/3 complex may induce the initiation of filopodia and thus regulate dendritic spine morphology (32).

Sigal et al. recently found that overexpression of PRG-3/LRP1 in HeLa and COS-7 cells increases the physiological number of filopodia independently of Cdc42 and ARP2/3 (58). Mainly expressed in the brain, PRG-3 transcripts where also found in liver, kidney and testis even through a much lower extend compared to PRG3 mRNA expression in the brain (51). In recent years, we have identified a further member of the PRG family, named plasticity-related gene-5 (PRG-5). Here, I showed that PRG-5 induces spine-like structure in young primary neurons.

2. The aim of work

PRG-1 and PRG-5 are both members of the plasticity-related genes (PRGs), a new class of transmembrane protein, mainly expressed in brain tissue. In my thesis, I focused my attention in two distinct aspects of these genes.

I investigated the regulatory elements of the upstream region of PRG-1 and then, I characterized the new member of PRG family, PRG-5, able to induce spine structures in young primary neurons.

PRG-1: the upstream region of PRG-1, a DNA chromosomal region before the transductional start point (ATG), was amplified by PCR and cloned in a reporter expression vector, pGL-3 luciferase basic vector, to analyze the putative „proximal promoter" responsible of the activation and neuron specific expression of PRG-1

PRG-5: I analyzed the putative protein structure and the mRNA expression level of PRG-5 in different mouse tissues, cellular types and developmental stages of mouse. Afterwards, I investigated the intracellular localization of PRG-5 in neurons cultivated at least one day in vitro and its ability to induce spine like structures in these young primary neurons.

Finally, I showed, by mutagenesis experiments, the amino acids of PRG-5 directly involved in the spine-induction process.

3. Material and Methods

3.1 Material

3.1.1 Laboratory equipment

Balance	Mettler-Toledo GmbH, Giessen
Centrifuges	RC-M150 GX, RC-5B, Ultra Pro80, Sorvall GmbH, Bad Hamburg
Dual Luciferase system	FB12 Luminometer, Bethold
Filmprocessor	Curix 60, Agfa-Gevaert, NV, Mortsel, Belgium
Gel filtration column	Amersham Pharmacia Biotech Europe GmbH, Freiburg
Gel electrophoresis equipment	Bio-Rad Laboratories GmbH, München
GeneAmp PCR system 9600	PE Biosystems, The Perkin Elmer Corporation, CA, USA
Heating block	Grant Instruments, Cambridge, UK
Incubator for HEK-293 and neuron cells	Memmert, Germany
Incubator shaker	New Brunswick scientific GmbH, Nürtingen
Microscope	Eppendorf AG, Hamburg
Confocal microscope	Leica TCS SP5, Microsystem, USA
PCR	Mx3005P Real Time Stratagene, La Jolla, CA, USA System
Pipettes	Eppendorf AG, Hamburg
pH meter	Mettler-Toledo GmbH, Giessen
Spectrophotometer	Bio-Rad Smart sectm3000, BioRad Laboratories GmbH, München
Thermocycler	PTC100, MJ Research, Inc.; Watertown, USA Biometra, Biometra GmbH, Germany Bior-Rad Laboratories GmbH, München
Vertical gel electrophoresis system	Bio-Rad Mini Protean II

UV-Transiluminator	Biometra, Götttingen
Vortex	Vortex Genie 2, Bender und Hobein AG, Zürich, Switzerland
Water bath	GFL, Germany
Western blot apparatus	Bio-rad Transblot SD apparatus

3.1.2 Solutions and buffers.

Antibiotics (1000x)	50mg/ml Ampicillin
	30mg/ml Kanamycin
LB medium	10g Bacto-tryptone
	10g NaCl
LB agar	LB medium
	15g/l bacto agar
6x DNA loading buffer	0.2% bromophenol blue
	60% Glycerol
	60mM EDTA
5x RNA loading buffer	50% Glycerl
	1mM EDTA
	0.25% Bromophenol blue
50x TAE buffer	242g Tris base
	57.1 ml Glacial acetic acid
	100 ml 0.5M EDTA pH8.0
10x TBS buffer	100mM tris
	9% NaCl
	pH adjusted to 7.4
PBS buffer	137mM NaCl
	2.7mM KCl
	4.3mM Na_2HPO_4

	1.47mM KH$_2$PO$_4$
	pH adjusted to 7.4
Blocking buffer	1x TBS
	3% BSA
Western blotting running buffer	20mM Tris
	150mM Gylcin
	20% Methanol
	0.08% SDS
Ponceau-S staining solution 0.2%	Ponceau
	3% Trichloroacetic acid
SDS-PAGE	0.20g SDS
	1.5M Tris-HCl pH8.8
	50 ml H$_2$O
3% Acrylamide pre-mix	3.8 ml 40% Acrylamide 29:1
	0.5M Tris-HCl pH 6.9
	10 ml SDS
	6 ml 50% sucrose
	500 µl 1% Bromophenol Blue
	50 ml H$_2$O
8% Resolving gel	2 ml 40% Acrylamide 29:1
	1.5M Tris-HCL pH8.8
	1.7 ml SDS
	1.3 ml 50% Sucrose
	43 µl 10% Ammonium persulfate
	9 µl TEMED
	4.948 ml H$_2$O
Stacking gel	4 ml 3% Acrylmaide pre-mix
	28 µl 10% Ammonium persulfate
	6 µl TEMED (stir quickly)

2x SDS gel-loading buffer	100mM Tris-HCl pH 6.8
	4% (w/v) SDS
	0.2% (w/v) bromophenol blue
	20% Glycerol

200 mM β-Mercaptoethanol (added just before the buffer was used)

Protein elution buffer	25mM Tris
	192mM Glycine
	0.1% SDS

Glycerol stocks plasmids clones were prepared in 25% glycerol and stored at -80°C.

3.1.3 Biological substances and Kits

Bacteria	DH5α chemocompetent cells, Promega, Mannheim
	XL-Gold chemocompetent cells, Stratagene, CA, USA
	BL-21 (RIL) bacterial expression cells, Stratagene, CA, USA
HEK-293 cells	(ATCC, Manassas, VA, USA)
Antibiodes	Moleculare Probes, Invitrogen Santa Cruz
	Biotechnology Inc., CA, USA AbCam, USA
	Molecular Probes, Invitrogen Sigma, USA Sigma Aldrich - Germany
Mice	C57/B16, Germany
Enzymes and reagents	DNaseI, RNase-free, RNaseH, Roche Diagnostics, Mannheim and Promega Madison, USA
	Taq DNA Polymerase, Promega, Madison, USA
	Taq DNA Herculase, Stratagene, La Jolla, CA, USA
	TaqMan Universal PCR Master Mix, Applied Biosystems, USA T4 DNA ligase, BioLabs, England

	Restriction enzymes, BioLads, England Ethidium bromide, Sigma, Deisenhofen Agarose, Roth, Karlsruhe
	Hepes, Roth, Karlsruhe
	pcDNA3.1/zeo(+) vector, Promega, Madison, USA pGL3basic vector, Promega, Madison, WI, USA pEGFP vector, Promega, Madison, USA
	Fugene6-reagent, Roche, Mannheim, Germany
	Effectene, Qiagen, Hilden, Germany
Gel elution and PCR	Nucleospin Extract kit, Macherey-Nagel
	Düren clean-up kits
Gel purification Kit	Qiagen, Hilden
RT-PCR	Qiagen, Hilden
TopoTA-cloning Kit	Promega, Mannheim, Germany
Multi site mutation Kit	QuikChange, Stratagene, CA, USA
Protein fraction	PromoKine Cell Fractionation, Heidelberg, Germany

3.1.4 Software

MetaMorph Systemss 6.2r4	Universal Imaging, Downingtown, PA
Leica Confocal Software v. 2.61	Leica
Adobe Photoshop 6.0	Adobe Systems Incorporated, San Jose, CA, USA DNASIS MAX Hitachi Software Engineering, CA, USA
Graphpad – Prism5	Statistical analysis, USA software program.
ExPASy Compute pI/MW	http://www.ic.sunysb.edu/stu/shilin/rnai.html tool
Blast	http://blast.ncbi.nlm.nih.gov/Blast.cgi

3.2 Methods

3.2.1 Vector construction of PRG-1

To generate promoter deletion constructs of PRG-1, I amplified specific-length fragments from the human and mouse total cDNA with Herculase (Stratagene, La Jolla, CA, USA). Amplified fragments were restricted and inserted into pGL3basic (Promega, Madison, WI, USA). Denomination of vectors referred to the position of the inserted region in relation to the translation start point, ATG.

3.2.2 Restriction enzymes

New England Biolabs (NEB) provided the restriction enzymes used in this study. The restriction enzyme digests were performed on at least 1µg DNA in 1X the final recommended buffer suggested by the manufacturer and using at least 1 unit of the specified enzymes. The reactions were incubated at 37°C for 1 hour for complete restriction enzyme digestion.

3.2.3 Agarose gel electrophoresis

Agarose concentrations ranging from 0.7% to 1% in TAE buffer (40 mM Tris-acetate, 1 mM EDTA containing 1 µg/ml ethidium bromide (EtBR; Sigma)), was used to cast agarose gels. DNA samples were loaded onto the gels with gel loading buffer (0.042% (w/v) bromophenol blue, 6.67% (w/v) sucrose) along with the DNA markers, λDNA/*Pst* I fragment markers (Invitrogen). Gel electrophoresis was performed between 90 to 110 V in TAE buffer for varying times to obtain optimal resolution. A gel documentation system (Bio-Rad) was used to visualize the gels by UV illumination.

3.2.4 Purifying linear DNA fragments

Linear DNA fragments were isolated from DNA agarose gels and purified using Qiagen spin columns. Based upon the fragment size or the amount of sample, electroelution was sometimes performed to elute the DNA from the gel. Elutip-D columns (Schleicher & Schuell) were used to further purify and concentrate the electroeluted DNA sample according to manufacturer's protocol, using 200 mM NaCl, 20 mM Tris-Hydrochloric acid (HCl) and 1.0 mM EDTA, pH 7.4.

3.2.5 Isolation of plasmids: mini prep

A small-scale preparation using an alkali lysis method was performed to screen colonies after transformation for positive recombinants. Single isolated colonies were selected and cultured overnight in LB containing the appropriate antibiotic. The next day, the cultured cells were resuspended in 50 mM glucose, 10 mM EDTA, 25 mM Tris- HCl (pH 8.0), 2 mg/ml lysozyme (Sigma) then lysed with a 200 mM NaOH, 1% sodium dodecyl sulfate (SDS) solution followed by the addition of 3 M sodium acetate, pH 5.2.
This precipitated out the bacteria chromosomal DNA, cellular protein and debris. The supernatant was extracted twice with phenol:chloroform:isoamyl alcohol followed by ethanol precipitation. The precipitated DNA was resuspended in TE (pH 8.0) containing 40 µg/ml RNAse A. DNA concentration was measured by standard A260/A280 spectrophotometric readings and visualized on agarose gels.

3.2.6 Bacterial cells

The bacterial host for vector propagation routinely used was DH5α *E. coli* cells from Promega. *E. coli* XL-Gold chemocompetent cells and BL-21 (RIL) bacterial expression cells were both obtained from Stratagene. These cells were grown at 37°C in a shaking incubator in a flask containing Lauria-Bertani (LB) broth consisting of 1.0% (w/v) tryptone (Difco), 0.5% (w/v) yeast extract (Difco) and 1.0% (w/v) NaCl supplemented with the appropriate antibiotic such as ampicillin (100µg/ml) or kanamycin (50 µg/ml), depending on the vector. For isolation of transformed bacterial cells, the cells were plated onto selective LB-agar plates, containing LB broth with 1.5% (w/v) agar supplemented with the appropriate antibiotic, and incubated inverted at 37°C overnight.

3.2.7 Transient transfection & dual luciferase assay

All cells were transiently transfected with equal amounts of the various luciferase-fusion constructs or pGL3basic in serum-free growth media using transfection reagents according to the manufacturer's protocol. pHRLtK (Promega, Madison, WI, USA) containing the *Renilla* luciferase gene fused to the herpes simplex virus thymidin kinase promoter was cotransfected in order to normalize transfection efficiencies. HEK-293 cells (ATCC, Manassas, VA, USA) were transfected with Fugene6-reagent (Roche, Mannheim, Germany). Primary neurons and astrocytes were transfected with Effectene (Qiagen, Hilden, Germany). The cells were lysed 48 hours after transfection and the luciferase activities of the cell lysates were determined using the dual-luciferase reporter assay system (Promega, Madions, WI, USA). Transfections with each construct were performer at least three times.

3.2.8 Statistical analysis

Statistical analysis was performed using the Mann Whitney U-test, which was performer using Prisms 5 software (Graphpad Software, San Diego, USA). Data are expressed as means +/- standard deviation. A critical value for significante of $p<0.05$ was used.

3.2.9 RNA isolation

Organs were dissected from mice and homogenized in TRIzol reagent. Primary neurons, microglial cells, or astrocytes were scraped in 1 x PBS, centrifuged for 5 min at 900 rpm and

4°C. Cell pellets were dissolved in 1 ml TRIzol reagent. Total RNA was isolated according to the TRIzol protocol (Invitrogen). Following precipitation and drying, RNA was resuspended in RNase und DNase-free water, an aliquot quantified by A_{260nm} spectrophotometry (Biomate 3 spectrometer, Fisher Scientific), and stored at −80 C.

3.2.10 DNAse treatment of total RNA sample and RNA cleanup

DNAse treatment was used to remove DNA from total RNA preparation. To the total RNA sample (~ 2 µg/µl diluted in RNAse-free water) 5 µl 10x RQ1 DNAse buffer, 5 µl 100 µM DTT and 2 units RQ1 RNAse-free DNAse (Promega) was added to a final volume of 50 µl. The mix was incubated for 30 min at 37°C. After the enzymatic reaction the volume of the sample was adjusted to 100 µl and the RNAeasy kit (QIAGEN) was used to clean up the total RNA. The RNA was eluted twice in the same tube with RNAse-free water to obtain a higher total RNA amount.

3.2.11 cDNA synthesis

cDNA was synthesized using the High-Capacity cDNA Archive Kit (Applied Biosystems) according to the manufacturer's protocol. Briefly, 5 µg total RNA was reverse transcribed in a 50µl reaction containing 5 µl 10 x random hexamer primers, 2 µl 25 x dNTP mix and 125 U MultiScribe reverse transcriptase. cDNA was diluted 1:5 with RNase and DNase-free water and stored at −20 C. The quality of the amplified cDNA was controlled by β-actin PCR.

3.2.12 PCR

DNA was amplified by PCR using the high fidelity *Pfu* polymerase or Herculase enzyme. The reaction volume was 25µl.

Template DNA	30 ng
10x buffer	5 µl
dNTP's (10mM)	1 µl
Forward Primer (5 pmol	1 µl
Reverse Primer (5 pmol	1 µl
Pfu polymerase (3 U/µl	0.5 µl
H_2O	25 µl

The following PCR profile was used:

PRG-5-Flag

1. Initial denaturation   95°C 2 min
2. Denaturation   95°C 45 s
3. Annealing   64°C 30 s
4. Extension   72°C 1 min for every 1000 bps
5. 30 cycles
6. Final extension   72°C 10 min

pGL-3-PRG-1

1. Initina denaturation   95°C 5 min
2. Denaturation   95°C 1 min
3. Annealing   66°C 45 s
4. Extension   72°C 1 min for every 1000 bps
5. 30 cycles
6. Final extension   72°C 10 min

The PCR products were agarose gel extracted.

3.2.13 Molecular cloning of PRG-5 and sequence analysis

Vector construction of pcDNA3.1Zeo(+)/PRG-Flag involved ligation of the gel-purified 0.9 kb of PRG-5-Flag fragment (*EcoI/NotI* cut site) into the pcDNA3.1Zeo(+) vector, resulting in the 6.4 kb vector (Fig. 2.1).

PRG-5 full-length clones were amplified by reverse transcription-polymerase chain reaction from postnatal mouse cDNA. Database searches used BLAST on the website of the National Center for Biotechnology Information. Multiple sequence alignments (GenBank Accession no. ACJ60628, human PRG-5; AAS80161, mouse PRG-5; NP_001101190, rat PRG-5) were made using DNAsis Max v2.0 software (Hitachi, Olivet Cedex, France). Transmembrane structure was predicted with ProDom and Swissprot databases. cDNA construct of mouse PRG-5, containing a C-terminal Flag, were used for the transiently transfection of HEK-293 cells and primary neurons cells. Briefly, cDNA from mouse brain was amplified by PCR using primers:

For_EcoRI: 5' – gaa ttc atg ccc ctg ctg ccc gtg gcg ctc atc agc –
3' Rev_NotI: 5' – gcg gcc gct cag tcg tca tcg tct ttg tag tct gtg act tcc gca aag gca gtg acg tgg ttc tgc ag – 3'

Reverse primer contains the Flag sequence. The oligonucleotides were synthesized by Metabion (Munich, Germany). The PCR products were purified by column (Qiagen), sub-clone in TOPA-Ta vector (Invistrogen), checked the sequence and then cloned in pCDNA3.1(+)zeo expression vector (Invitrogen) by EcoRI and NotI.

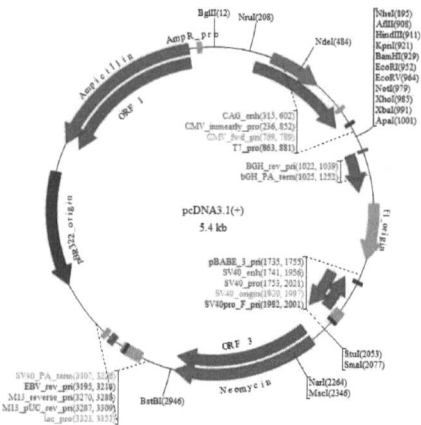

Fig. 3.1 Diagram of pcDNA3.1Zeo(+) vector

3.2.14 Ligation and transformation

Purified PCR products were cloned in a total volume of 25 µl ligation reaction as below. This reaction mix was incubated at room temperature for 1 hour.

10X buffer	2.5 µl
Vector	5 µl
Insert	15 µl
T4-ligase	1 µl
H$_2$O	1.5 µl

For transformation, 16 µl of the ligation reaction were added to 750 µl thawed competent bacteria and incubated on ice for 10 minutes. Then, bacteria were heat shocked for 30 seconds at 42°C and put on ice 2 minutes. After that, 750 μl SOC media was added and incubated at 37°C for 1 hour with 800 rpm shaking.

3.2.15 Clone analysis

Minipreps carried out extraction of the plasmids of transformed bacterial cells. The restriction enzyme digestion reaction mix contained:

10Xbuffer	2μl
Restriction enzyme (10U/μl)	1μl
DNA	1μl
H_2O	16μl

Bacterial clones shown to contain the target plasmids were preserved for further use. Sterile glycerol was added in a ratio of 1:3 and aliqouted in 1.5 ml cryotubes then stored at –80°C.

3.2.16 Site direct mutants of mouse PRG-5

We have synthesised two complimentary oligonucleotides, containing the nucleotides we had mutated (uppercase) and generated mutants of mouse PRG-5 using the QuikChange protocol (Stratagene).

F_Mut_S193W:	5' – cgg aaa aca ttc cca TGT aag gaa gct gcc ctg – 3'
R_Mut_S193W:	5' – cag ggc agc ttc ctt CCA tgg gaa tgt ttt ccg – 3'
F_Mut_E195G:	5' – aca ttc cca tcc aag GGC gct gcc ctg agt gtc – 3'
R_Mut_E195G:	5' – gac act cag ggc agc GCC ctt gga tgg gaa tgt – 3'
F_Mut_E195H:	5' – aca ttc cca tcc aag CAC gct gcc ctg agt gtc – 3'
R_Mut_E195H:	5' - gac act cag ggc agc GTG ctt gga tgg gaa tgt – 3'

F_Mut_R241E: 5' - ctt act gga ctc aac GAG gta gcg gaa tat cga – 3'
R_Mut_R241E: 5' - tcg ata ttc cgc tac CTC gtt gag tcc agt aag – 3'

Briefly, PRG-5 was amplified by PCR using the high fidelity *Pfu* Turbo DNA and mutated primers in a final reaction volume of 50μl:

Template DNA	30 ng
10x buffer	5 μl
dNTP's (10mM)	1 μl
F_Mut Primer (125ng)	1 μl
R_Mut Primer (125ng)	1 μl
Pfu polymerase (2.5U/μl)	1 μl
H$_2$O	38 μl

The following PCR profile was used:

1. Initial denaturation	95°C	30 s
2. Denaturation	95°C	45 s
3. Annealing	55°C	1 min
4. Extension	68°C	1 min for every 1000 bps
5. 18 cycles		
6. Storage	4°C	hold

We added 1 μl of the *Dpn* I restriction enzyme (10 U/μl) directly to each amplification reaction and incubated each reaction at 37°C for 1 hour to digest the parental (i.e., the nonmutated) supercoiled dsDNA before transfection in E.Coli.

3.2.17 HEK-293, preparation and transfection

HEK-293 cells, containing the adenoviral E1 genes transformed into the cells, were purchased from Microbix (ATCC, Manassas, VA, USA) and grown in Minimum Essential Medium containing Earle's Salts (EMEM; Gibco) supplemented with 10% FBS and gentamicin solution. The cell lines were mostly passaged using Trypsin/ethylenediamine tetracetic acid (EDTA; Gibco) whereas HEK-293 cells were passaged using a 1X citric saline solution (10% (w/v) KCl and 4.4% (w/v) sodium citrate). All cells lines were cultured in a 37°C humidified CO_2 incubator with a 5% CO_2 atmosphere. Trypan-Blue (Gibco) was used for cell counting on a haemocytometer.

HEK-293 cells were routinely maintained at 37°C with 5% CO_2 in Dulbecco's modified Eagle's medium (DMEM) supplemented with 10% fetal bovine serum (Gibco), 100 U/ml penicilin and 100μg/mL streptomycin, respectively as previously described (7) into 12-well plates or dishes with cell density of 10,000 – 20,000 cells/cm^2 and transfected with calcium; in two separate tubes I prepared;

solution A:

DNA	1-2 μg
$CaCl_2$ 2M	12,3 μl
H_2O to	100 μl

solution B:
HBS 2x	100μl

Add the 100μl solution A to the solution B (100 μl) and vortex for a few seconds. Incubate at RT for a 3-5 min, mix again and dropwise onto the media.

3.2.18 Primary neurons, preparation and transfection

Primary neurons were isolated from C57BL/6 mouse embryos at embryonic day 18. Brain from several embryos were collected and washed twice in ice-cold HBSS (Invitrogen). The tissue was incubated in 4 ml HBSS and 400µl trypsin (Invitrogen) for 15 min at 37°C, resuspended in plating medium (MEM supplemented with 10% horse serum, Invitrogen, 0.6% glucose, 100 U/ml penicillin and 100µg/ml streptomycin). After dissociation, the neurons were plated in plating medium onto poly-L-lysine (Sigma)-coated cover slips into 12-well plates with cell density of 100.000 cells/cm^2. Three hours after plating, cells were washed twice with phosphate-duffered saline (1 x PBS) and incubated in Neurobasal A medium (Invitrogen) supplemented with 2% B27 (Invitrogen), 0.5 mM glutamine, 100 U/ml penicillin, 100 µg/ml streptomycin at 37°C and 5% CO_2. Primary neurons were routinely maintained at 37°C with 5% CO_2 in Neurobasal A medium (Invitrogen) supplemented as described above for 1 or 3 days in vitro and transfected using efectine (Qiagen).

3.2.19 qRT-PCR

Each PCR reaction contained: 8 µl H_2O, 10 µl TaqMan® Universal PCR Master Mix (Applied Biosystems), 1 µl cDNA and 1 µl TaqMan Gene Expression Assays: for PRG5 (Assay ID Mm01310525_m1), GAPDH (glyceraldehyde-3-phosphate Dehydrogenase; Assay ID 4352932E) and β-actin (Assay ID 4352933E). For HPRT (hypoxanthine phosphoribosyltransferase), separate primer and probe were used primer mix:

    For: 5' – atc att atg ccg agg att tgg aa – 3'
    Rev: 5' – ttg agc aca cag agg gcc a – 3'

And probe:

    5' – tgg aca gga ctg aaa gac ttg ctc gag atg – 3'

The reactions were run on the ABI PRISM™ 7700 Sequence Detection System (Applied Biosystems). Standard curves were produced with serial dilutions of cDNA from P5 mouse brain with amplification efficiency between 90 and 100%. Each result is the average of at least three separate experiments.

3.2.20 Immunocytochemistry

Primary neurons and HEK-293 cells, were fixed, after 24 hour from the transfection, in 4% paraformaldehyde for 20 min and permeabilized with 0.2% Triton X-100 (Sigma) in PBS for 5 min, washed three times with PBS 1X and incubate with block-solution, containing 10% FCS in PBS, over night at 4°C. Cells were incubated with first antibodies 2 hours at room temperature and washed 3 times with PBS 1X. Second antibodies applied 1 hour at room temperature. After 3 more washes with PBS 1X, glasses were mounted in FluoroSave Reagent. Confocal images were processed with Leica software.

3.2.21 In situ hybridization

For hybridization, I used antisense oligonucleotides:
   5' – ggc agt gtg gtt ctt ttc cca agg gct ttc tact cg agg aat gtt gg – 3'

Complementary to the mouse sequence of PRG-5, the oligonucleotide was synthesized by Metabion (Munich, Germany). The specificity was confirmed by a BLAST GenBank search to rule out cross hybridization with other genes. Horizontal cryostat sections (20µm) were fixed in 4% paraformaldehyde (PFA), washed in 0.1M phosphate buffered saline (PB) (pH7.4) and dehydrated through an ascending series of pure ethanol. The oligonucleotides were end-labelled using terminal desoxynucleotide transferase (Boehringer Mannheim, Germany) and x-dATP (DuPont NEN). Probe labeling was performed for 10 min at 37°C. The radioactive probes were purified using BioSpin6 Chromatography Columns (Bio-Rad). I used 100.000 – 200.000 cpm labeled oligonucleotides in 50µl hybridization buffer (50% formamide, 10mM Tris-HCl, pH 8.0, 10 mM PB, pH 7.2, 2x SSC, 5mM EDTA, pH 8.0, 10% dextran sulphate, 10 mM dithiothreitol (DTT), 1mM β-mercaptoethanol and 200 ng/µl tRNA) per section. Hybridization was performed for 16h at 42°C in a humidified chamber, after which the slides were washed as follows: 1 x 60 min in 0.1 x SSC at 56°C and 1 x 5 min in 0.05 x SSC at room temperature. Finally, the sections were rinsed in H_2O at room temperature and dehydrated in 50, 75 and 96% ethanol. For autoradiography, slides were exposed to Kodak X-OMAT AR X-ray films (Amersham, Heidelberg, Germany) for 15 days. No signals were detected on sections hybridized with specific antisense oligonucleotide when unlabeled oligonucleotide was added in 100-fold surplus or with sense probes of the respective oligonucleotides. Following exposure, sections were rehydrated using a decreasing ethanol series, washed in phosphate-buffered saline, Nissl counterstained (cresyl violet acetate; Sigma, Germany), dehydrated through an ascending series of ethanol, flat-embedded with Entellan (Merck-Germany) and cover-slipped. Sections were then digitally photographed (Zeiss, Axioplan, Germany).

3.2.22 SDS-polyacrylamide-gel electrophoresis (Protein)

Protein samples (prepared in Laemmli buffer followed by boiling at 95°C for 5 min) were separated on 10-12% mini SDS-PAGE gels (Mini-protein II Dual Slab Cell, Bio-Rad; Laemmli 30 3. Methods 1970). In this system, proteins denatured in the presence of SDS and 2-mercaptoethanol as thiol reducing agent acquired a rod-like shape and a uniform charge-to-mass ratio proportional to their molecular weights. The gels were stained with colloidal Coomassie stain and the protein sizes were determined by comparing the migration of the protein band to a molecular mass standard (PageRuler prestained Marker, Fermentas).

3.2.23 Western blot

Protein concentration was determined by BCA protein assay (Pierce, IL, USA) using BAS as a standard. Samples were separated by 10% SDS-PAGE and the proteins were transferred to a PVDF membrane /Amersham Biosciences, CA, USA). The membrane was incubated with primary antibody PAP2D (AbCam, UK) 1:700 or Flag antibody (Sigma) 1:1000, followed by horseradish-peroxidase-conjugated secondary antibody. The ummunoreactive bands were detected with SuperSignal West Pico chemiluminescent substrate (Pierce). Subcellular protein fractions (cytosol and membrane) were prepared using the kit PromoKine Cell Fractionation Kit protocol (PromoKine, Germany).

4. Results

4.1 Identification of proximal promoter of PRG-1

In order to elucidate the peculiar expression profile of PRG-1 in neurons, a reporter mouse model expressing YPF controlled by the native PRG-1 promoter, was studied (21a). PRG-1 was able to show that 200 kbp sequence enclosing the PRG-1 gene is effectual to confer cell type specific expression. By RLM-Race in rat as well as in mouse, and confirmed these via PCR-analysis, it were mapped additional transcription start points (TSS) In relation to the translation start point ATG = +1, PRG-1 mapped several transcription start sites (TSS) in rat and mouse. Rat transcripts start at position: -201, -160, -156, -96, +46, +69 while mouse transcripts start at position: -149, -144, -101, +2, +43, +45, +71 and +79 (21a) (Fig. 4.1).

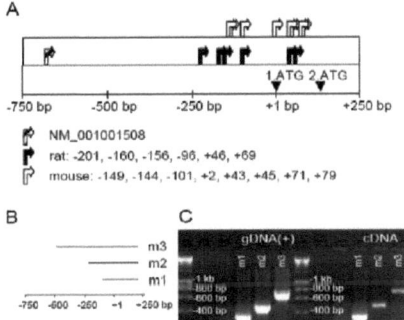

Fig. 4.1 PRG-1 is transcribed from multiple transcription start points in rodents. (A) Mapped transcription start sites of PRG-1 in rat (filled arrows): -201, -160, -156, -96, +46 and +69. The positions are designated in relation to the first translation start site (ATG=+1). Mapped transcription start sites in mouse (open arrows): -149, -144, -101, +2, +43, +45, +71 and +79. A second putative translation start site is conserved in mammals, in mouse and rat at position +145, in the human sequence at position +143. (B) Position of amplificates in relation to the first murine ATG. (C) Positive control amplified from mouse genomic DNA (gDNA) for comparison of amplification efficacy. PCR products were amplified from mouse cDNA.

Some TSS were located downstream of the first translation start point, revealing a second putative translation start site, which is conserved in the mammalian PRG-1 sequences. In mouse and rat, the second ATG was located at position +148, and in human at position +145.

To further investigate the core promoter of mouse PRG-1, I performed dual-luciferase-assays with several deletion-constructs. The mouse 1869-bp fragment (m-1724/+145) was subcloned into the promoter-less luciferase reporter vector, pGL3, in sense and antisense orientations, and its promoter activity was analyzed in primary rat neurons and astrocytes. As shown in Fig. 4.2, luciferase activity was minimal, in both primary neurons and astrocytes, when expression was driven by the 1869-bp sense and antisense fragment respectively (m-1724/+145 and m-1724/+145rev), compared with the pGL3 vector.

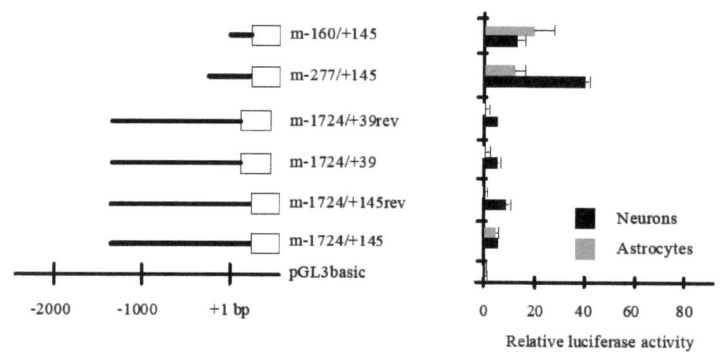

Fig. 4.2 Identification of a sequence element in the mouse PRG-1 promoter sufficient to confer neuronal transcription. PRG-1 promoter 5' end deletion constructs were ligated in the dual luciferase reporter gene vector pGL3. constructs and expression control pHRLtk (promega) were co-transfected in primary rat neurons and astrocytes. The relative luciferase activity is defined as ratio of Firefly luciferase versus Renilla luciferase. Basal-level luciferase activity of cells transfected with pGL basic and pHRLtk (promega) was taken as 1%. Results are expressed as fold increase +/-SD. The experiments were repeated at least three times independently. The positions of the construcuts were designated in relation to the first translation start site (ATG=+1).

Luciferase activity was 40 fold higher in neurons, respect astrocytes, when expression was driven by the 400-bp fragment (m-277/+145) (Fig. 4.2).

Transfection of human PRG-1 promoter deletion-constructs in primary rat neurons and astrocytes resulted, as well in an identification of a 450 bp fragment (h-300/+143). The luciferase activity was 50 fold greater than that of control plasmid in neurons, compared to astrocytes (Fig. 4.3). Truncation of this fragment to 300 bp (h-169/+143) leaded to a 50% reduction of specificity for neuronal transcription down to the level of stimulation in astrocytes. Transfection of this 300 bp-sequence in reverse orientation (h+143/-169) leaded to almost complete loss of transcriptional activity. All constructs with a 3'-extension to the second putative translation start site display enhanced transcriptional activity compared to controls. Interestingly, the addition of 5'-sequences to the 450 bp minimal promoter led to a strong decrease of transcriptional activity (Fig. 4.3).

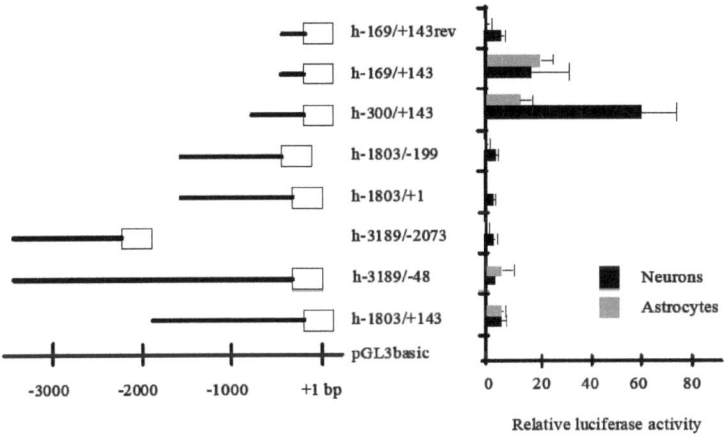

Fig. 4.3 Human PRG-1/pGL3 reporter constructs and expression control pHRLtk were co-transfected in primary mouse neurons and astrocytes.

4.2 Expression pattern of PRG-5

Human PRG-5 was firstly cloned as phosphatidic acid phosphatase type 2D (PAP2D) and the two isofroms were deribed as PAP2D-v1 and PAP2D_v2 (62). The two isoforms differ in 15 bp at the C-terminus and the five amino acids not presented in PAP2D_v2 are VTSVQ, corresponding, in mouse PRG-5, to the amino acids VTSLQ (Fig. 4.4).

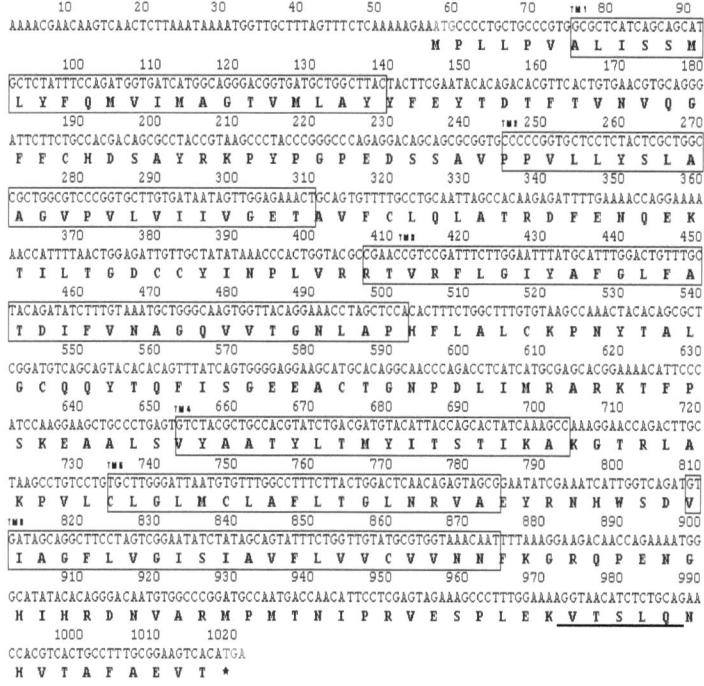

Fig. 4.4 Mouse PRG-5 nucleotide and amino acid sequence. Start and stop codons of mPRG-5 (GeneBank Accession no. AAS80161) are in red. Putative transmembrane domains (TM) are boxed in grey and were predicted with ProDom and Swissprot databases. The five amino acids not presented in PAP2D_v2 are underlined.

In my studies, I specifically investigated the longest isoform in mouse. Mouse PRG-5 full-length clones were amplified by reverse transcription polymerase chain reaction from postnatal and adult mouse hippocampus cDNA. The PCR product it's been subcloned in TA vector to check the correct sequence and then fused on the pCDNA3.1+vector (Fig. 4.5).

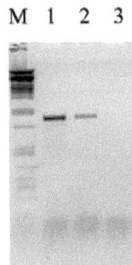

Fig. 4.5 PCR from total amount of mouse cDNA. M = Lambda marker digested by PstI. The single band presence in the lines 1 and 2 are PRG-5 amplified by PCR, using decries amount of mouse cDNA as template. Line 3 is a negative control.

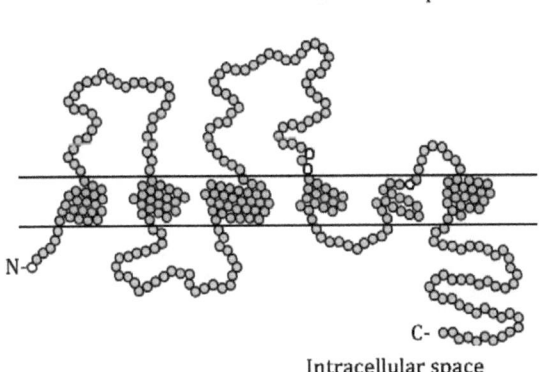

Fig. 4.6 Structure model based on the amino acid sequence of mouse PRG-5 (ExPasy). The double line represents the plasma membrane.

The protein contains 321 amino acids and its molecular weight measures approximately 35 KDa (Fig. 4.6). In silico analysis (ProDom and Swiss-Prot databases) of human, rat and mouse PRG-5 revealed protein homologies conserved up 90% (Fig. 4.7).

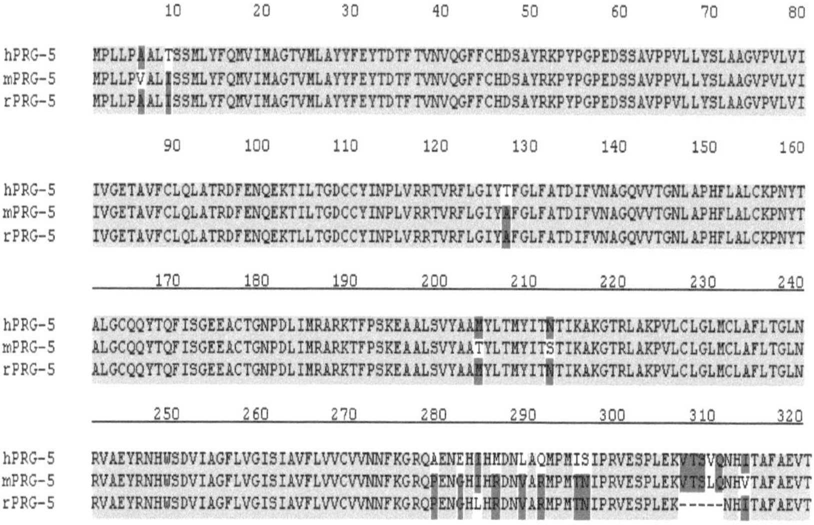

```
                10        20        30        40        50        60        70        80
hPRG-5  MPLLPAALTSSMLYFQMVIMAGTVMLAYYFEYTDTF TVNVQGFFCHDSAYRKPYPGPEDSSAVPPVLLYSLAAGVPVLVI
mPRG-5  MPLLPVALISSMLYFQMVIMAGTVMLAYYFEYTDTF TVNVQGFFCHDSAYRKPYPGPEDSSAVPPVLLYSLAAGVPVLVI
rPRG-5  MPLLPAALISSMLYFQMVIMAGTVMLAYYFEYTDTF TVNVQGFFCHDSAYRKPYPGPEDSSAVPPVLLYSLAAGVPVLVI

                90       100       110       120       130       140       150       160
hPRG-5  IVGETAVFCLQLATRDFENQEKTILTGDCCYINPLVRRTVRFLGIYTFGLFATDIFVNAGQVVTGNLAPHFLALCKPNYT
mPRG-5  IVGETAVFCLQLATRDFENQEKTILTGDCCYINPLVRRTVRFLGIYAFGLFATDIFVNAGQVVTGNLAPHFLALCKPNYT
rPRG-5  IVGETAVFCLQLATRDFENQEKTLLTGDCCYINPLVRRTVRFLGIYAFGLFATDIFVNAGQVVTGNLAPHFLALCKPNYT

                170       180       190       200       210       220       230       240
hPRG-5  ALGCQQYTQFISGEEACTGNPDLIMRARKTFPSKEAALSVYAAMYLTMYITNTIKAKGTRLAKPVLCLGLMCLAFLTGLN
mPRG-5  ALGCQQYTQFISGEEACTGNPDLIMRARKTFPSKEAALSVYAATYLTMYITSTIKAKGTRLAKPVLCLGLMCLAFLTGLN
rPRG-5  ALGCQQYTQFISGEEACTGNPDLIMRARKTFPSKEAALSVYAAMYLTMYITNTIKAKGTRLAKPVLCLGLMCLAFLTGLN

                250       260       270       280       290       300       310       320
hPRG-5  RVAEYRNHWSDVIAGFLVGISIAVFLVVCVVNNFKGRQAENEHIHMDNLAQMPMISIPRVESPLEKVTSVQNHITAFAEVT
mPRG-5  RVAEYRNHWSDVIAGFLVGISIAVFLVVCVVNNFKGRQPENGHIHRDNVARMPMTNIPRVESPLEKVTSLQNHVTAFAEVT
rPRG-5  RVAEYRNHWSDVIAGFLVGISIAVFLVVCVVNNFKGRQPENGHLHRDNVARMPMTNIPRVESPLEK-----NHITAFAEVT
```

Fig. 4.7 Human, mouse and rat PRG-5 sequence alignment. Full-length human and rat sequences were taken from GeneBank. The accession numbers are AAS80161 (mouse PRG-5), ACJ60628 (human PRG-5) and NP_001101190 (rat PRG-5). Identical sequences between all three different species are shaded in yellow and between two of three changes in green.

As a PRG/LRP family member, PRG-5 is a six membrane-spanning protein with a C- and N-terminus locked on the cytoplasmic site. Although sequence alignments between all PRGs/LRPs and LPP-1 revealed conserved amino acids, especially in the catalytic sequence motif within the three loops (Tab. 1), PRG-5 lacks several crucial residues for ecto-phosphatase activity (26, 61, 64, 74).

		C1		C2		C3
mPRG-1	(206)	PYFLTVCKP	(251)	SQH	(297)	TRITQYKNHPVD
mPRG-2	(158)	PFFLTVCKP	(203)	SQH	(249)	TQITQYRSHPVD
mPRG-3	(154)	PYFLTVCQP	(198)	SKH	(245)	NRVAEYRNHWSD
mPRG-4	(156)	PHFLSVCRP	(207)	CKD	(254)	VRVAEYRNHWSD
mPRG-5	(149)	PHFLALCKP	(193)	SKE	(240)	NRVAEYRNHWSD
mLPP-1	(128)	PHFLAICNP	(169)	SGH	(216)	SRVSDYKHHVSD

Tab. 1 The consensus phosphatase sequence motif between PRGs and LPP-1 of mouse. In red the crucial amino acids of LPP-1 (58).

Through northern blot analysis with Pan Oligo, it is known that PRG-5 is mainly expressed in the brain and weakly in testis (Fig. 4.8)

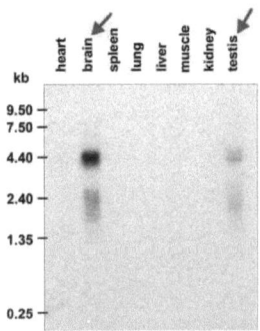

Fig. 4.8 Northern blot analysis of PRG-5 (11).

In situ hybridizations analysis highlighted tight regulation of mouse PRG-5 mRNA expression during brain development. PRG-5 was detected at embryonic day 17 (E17) and this expression is found in the hippocampal anlage, cortex and olfactory bulb. A strong hybridization signal was found in cortex and hippocampus at Post-natal day 0 and 5 (P0 and P5) (Fig. 4.9).

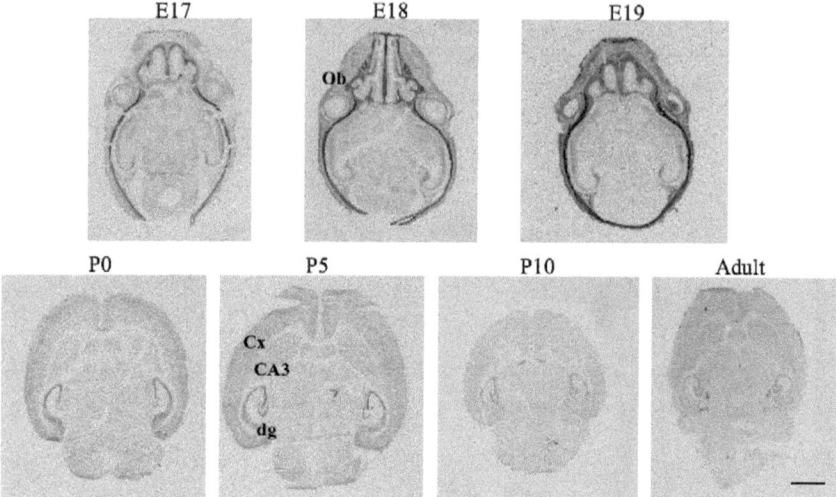

Fig. 4.9 PRG-5 expression during embryonic and postnatal development. At embryonic day 17 (E17) hybridization signal for PRG-5 can be detected in the hippocampal anlage and olfactory bulb. At perinatal stages (E19 – postnatal day 10 (P10)), a strong hybridization signal is found in the cortex and hippocampus as well in dentate gyrus (dg). Scale bars 1,8 mm. CA, cornu ammonis; cx, cortex; Ob: olfactory bulb.

After that, studies using mRNA expression of PRG-5, normalized with HPRT or GAPDH housekeeping gene, revealed that expression of PRG-5 mRNA were significantly higher in brain than all other tissues, confirming that its expression is indeed brain-specific (Fig. 4.10).

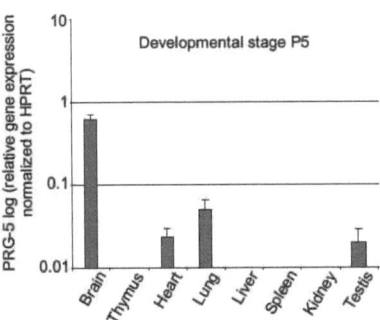

Fig. 4.10 Multi-tissue real-time PCR analysis of PRG-5 mRNA shows the brain mainly expression of PRG-5 mRNA at P5. HPRT was used as reference gene. Error bars are standard deviations.

PRG-5 is expressed as early as embryonic day 14 (E14), peaks at postnatal day 0 (P0) (Fig. 4.11) and is principally expressed in neurons ad opposed to astrocytes and microglia of the brain hippocampus (Fig. 4.12).

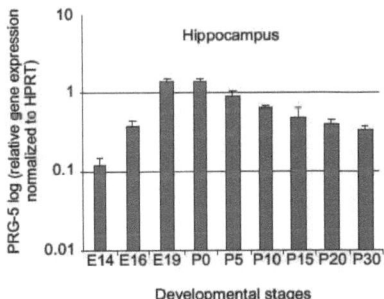

Fig. 4.11 Quantification of PRG-5 mRNA in hippocampus mouse by real-time PCR. PRG-5 mRNA expression in the hippocampus of developing mouse brain, normalized to HPRT. Error bars are standard deviations.

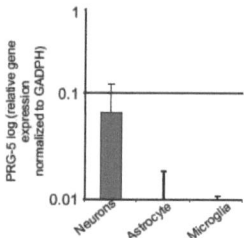

Fig. 4.12 PRG-5 mRNA expression in neurons, astrocytes and microglia, normalized to GAPDH. Error bars are standard deviations.

4.3 Sub-cellular localization of PRG-5.

By PCR technique using total cDNA of mouse brain at P0, we generated a PRG-5 flag protein (DYKDDDD), as a tag fused at the C-terminus, instead of an EGFP fusion protein. Indirect immunofluorescence analysis of fixed cells revealed that the protein was localized to both intracellular membrane structures and to the plasma membrane (Fig. 4.13).

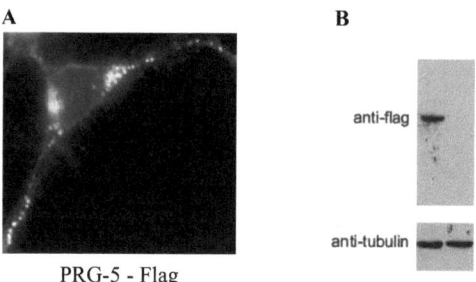

Fig. 4.13 (A) Overexpression of PRG-5 in HEK-293 cells. PRG-5 showed a punctuated distribution (white arrows) typically of the protein packed in vesicles. (B) Western blotting from total protein extraction from HEK-293. Asterisks indicate non-specific band.

In detail, intracellular localization of PRG-5 was analysed by confocal microscopy, using Golgi and Rab9 vesicles marker (Fig. 4.14).

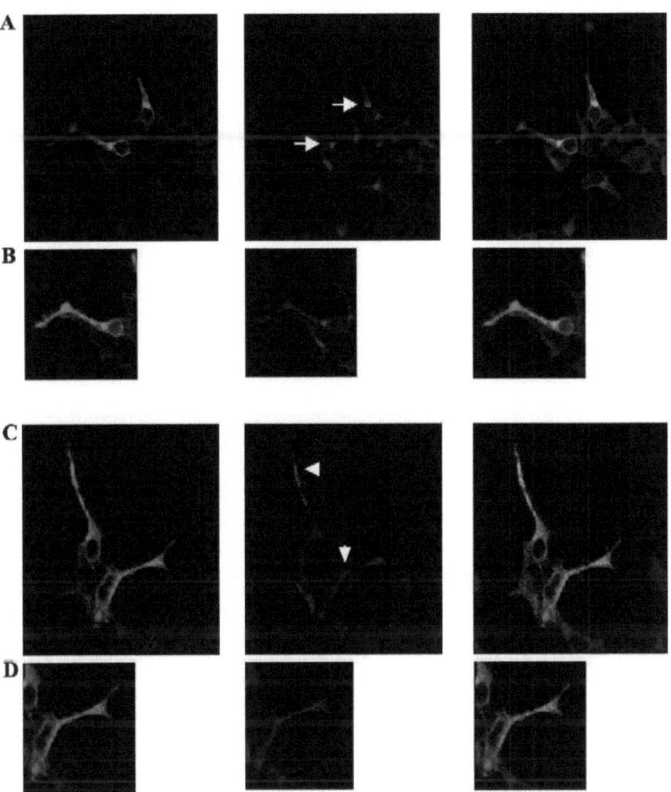

Fig 4.14 PRG-5-Flag co-localizes with Golgi and Rab9 – vesicles marker in HEK-293 cells. (A and C) HEK-293 cells were transfected with pCDNA3-PRG-5-Flag. Fixed 24 hours later and investigated by confocal microscopy. (A) PRG-5 co-localize with Golgi marker (white arrows) and (C) partially with Rab-9 (white arrowheads). (B and D) Higher magnification of A and C respectively. Scale bar are 10μm and 4μm.

4.4 PRG-5 induces spine formation in primary neurons.

In light of the brain-specific expression in mouse, especially in the case of neurons, we performed a series of overexpression experiments in mouse hippocampal primary, cultivated between DIV 2 and 4. At this age, PRG-5 was widely expressed. The mRNA expression of PRG-5 was revealed by RT-PCR. I used a couple of primer to amplified PRG-5 and β-actin as control. In line 1 and 2, as template, I used cDNA from total mRNA of primary neurons at DIV 2 and 4 respectively, in line 3 H_2O as negative control and in the line 4 a pcDNA-PRG-5 vector as positive control. The asterisk in line 4, respect the b-Actin primers, indicate an excess of primers and dNTPS used for the PCR (Fig. 4.15).

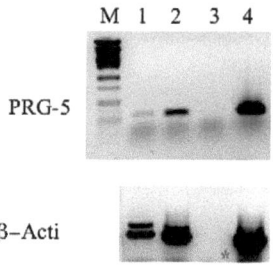

Fig. 4.15 RT-PCR from primary neurons at DIV 2 and 4 (line 1 and 2), H_2O (line 3) and pcDNA-PRG-5 (line4) with primers to amplify PRG-5. M = Lambda marker digested by PstI.

Western-blot analysis of protein extracted from neurons at different days *in vitro* (DIV) with a PAP2D antibody, specific for PRG-5, revealed that PRG-5 is expressed at both DIV2 and DIV4. In particular, PRG-5 was detected in the membrane of the neurons, but not in their cytosolic fractions (Fig. 4.16 DIV 2 and DIV 4). To study the overexpressed PRG-5 protein, I fused a Flag tag to the C- terminus of PRG-5. Membrane and cytosolic fractions from HEK-293 cells overexpressing PRG-5-Flag were tested using Western blot, and confirmed that both antibodies were able to detect PRG-5 specifically (Fig. 4.16 HEK-293).

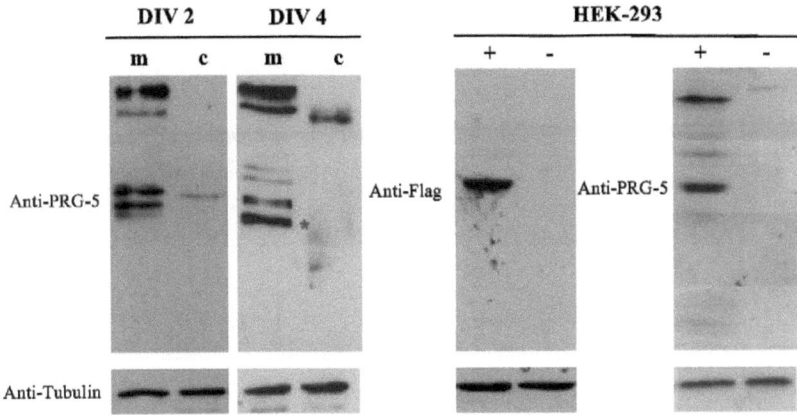

Fig. 4.16 Protein extraction was analysed by Western blotting using PAP2D (PRG-5) and Flag antibodies. Analysed membrane (m) and cytosolic (c) fractions showed endogenous expression of PRG-5 in primary neurons, cultivated 2 or 4 days in vitro. PRG-5-Flag overexpressed in HEK-293 cells was detected with Flag and PAP2D antibodies in the membrane fraction.

By means of confocal microscopy, a PRG-5-Flag signal was revealed on the body of the cell, in the proximal region of dendrites and at the distal neurites (Fig. 4.17). Higher magnification showed the presence of spine-like, mushroom-shaped and stubby structures along the plasma membrane. Distribution of PRG-5 along the plasma membrane appeared discontinuous, with a pronounced localization to many membrane protrusions (Fig. 4.17, white arrowhead).

The intracellular punctuated signal it was investigated in details using a α-clathrin vesicle, which is a specific vesicular carrier for membrane proteins. Clathrin spots were identified in the same area of PRG-5-Flag, suggesting a colocalization of PRG-5 with clathrin (Fig. 4.18, open arrows in central panel).

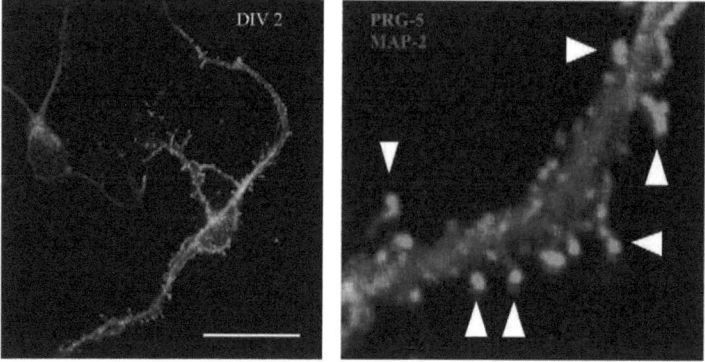

Fig. 4.17 Transfected primary neurons were fixed at DIV 2 and stained with antibodies to PRG-5-Flag (green) and MAP-2 (red). Scale bar represents 10 μm. Higher magnification of neuritis shows spine-like protrusions (white arrowheads). Scale bar represents 4 μm.

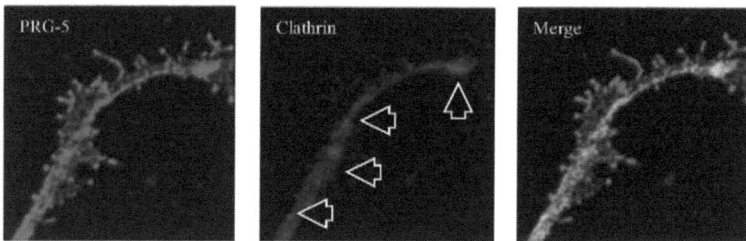

Fig. 4.18 PRG-5 (left panel) stained with antibodies to α-clathrin (central panel, open arrows). Merged panel shows PRG-5-induced spine structure and colocalization between PRG-5 and α-clathrin. Scale bar represents 3 μm.

More in-depth analysis of PRG-5 was performed on primary neurons at DIV 2 or 4 by means of overexpression using markers for actin and tubulin. It was possible to distinguish swollen structures with a massive presence of spine structures containing PRG-5 (Fig. 4.19 PRG-5-flag with actin and higher magnification, white arrowheads at the left panel). They originated from the plasma membrane, becoming more apparent along the processes. These protrusions were rich in polymerized actin and PRG-5-Flag was predominantly distributed uniformly along the thin neck and in the bulbous head of each single spine (Fig. 4.19, higher magnification, central and merged panel), whereas tubulin structures appeared only along the processes of the neurites but was not in the bulbous head (Fig. 4.19 PRG-5-flag with tubulin, white arrowheads). Primary neurons at DIV 4 overexpressing PRG-5-Flag showed the same distribution and shape of spines displayed in neurons at DIV 2 (Fig. 4.21). These types of spines were not found when EGFP alone was overexpressed in the same primary neurons (Fig 4.20 A and B and Fig. 4.21 B).

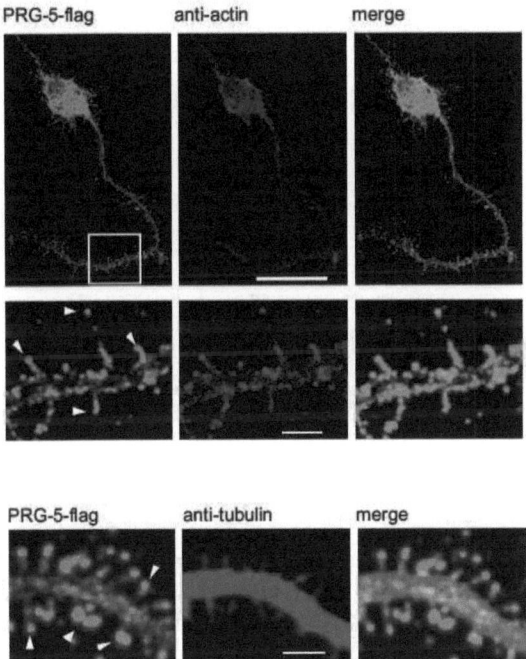

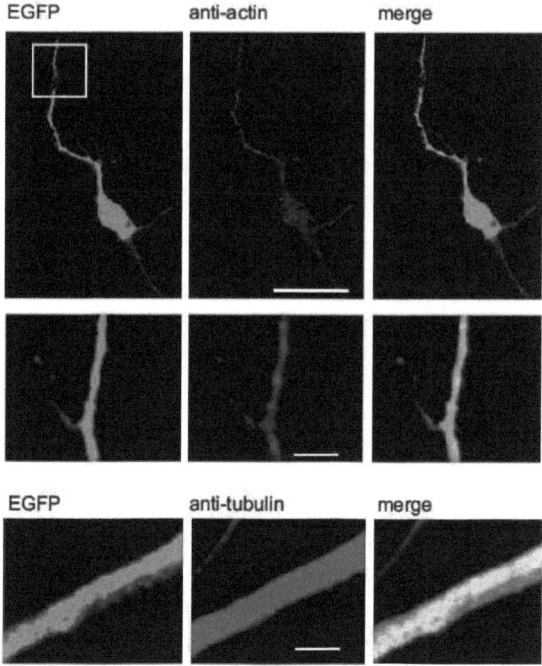

Fig. 4.20 Overexpression of EGFP did not alter the morphology of the cells or induce spine formations along the neuritis. The scale bar represents 10 μm and 4 μm for higher magnifications.

Fig. 4.19 PRG-5 induces the dendritic spine formation during the first stages of neuronal differentiation. Cultured primary neurons transfected with PRG-5-Flag (green) at DIV 2. Scale bar represent 10μm. High-magnification insets show that PRG-5 (green) induces mature, mushroom-shaped spines along individual neurites that extend through the peripheral region, visualized by phalloidin (red). Arrowheads indicate spines with PRG-5 in the bulbous head. Scale bar represents 4 μm. In higher magnification, primary neurons transfected with PRG-5-Flag (green) and visualized by tubulin (red). The white arrowheads indicate the heads of different spines induced by PRG-5. Scale bar represents 4μm.

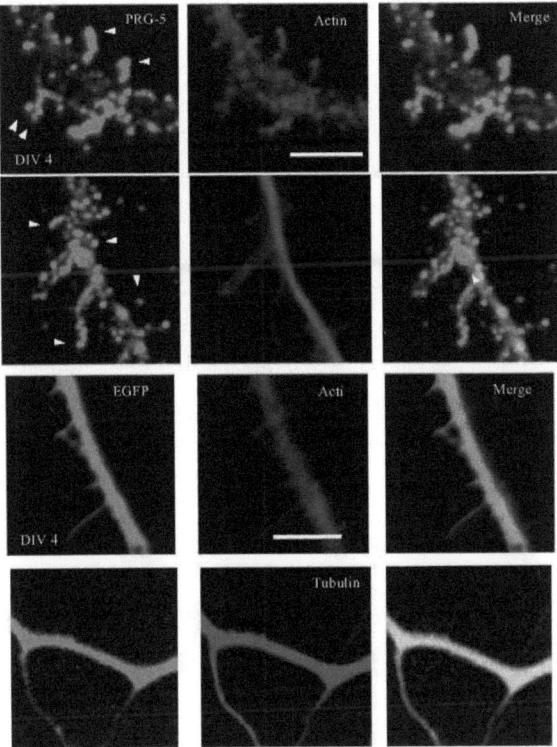

Fig. 4.21 Primary neurons transfected with PRG-5-Flag (green) at DIV 4 and visualized with actin (central bottom panels) or tubulin. Arrowheads indicate different mushroom-shaped and stubby spines. The scale bar represents 4 µm. EGFP overexpressed in primary neurons at DIV 4 and detected using actin (central bottom panels) or tubulin. The scale bar represents 4 µm.

4.5 Analysis of residues within the C1 – C3 domains of PRG-5

Sigal Y. J. et al. have reported that single exchange of the amino acids Ser198 and His200 within the C2 domain and Arg246 within the C3 domain of PRG-3 in COS7 and HeLa cells is crucial for a correct functioning of the protein and resulted in an overproduction of the number of filopodia (58). However, despite the fact that analysis of the sequence homology between PRG-5 and PRG-3 showed a similarity higher than 85% (GeneBank), the residues within the C1-C3 domains, responsible for filopodia induction in PRG-3, were not completely conserved in PRG-5 (Tab. 1). In fact, although the amino acids Ser198 and Arg246 within the C2 and C3 domains, respectively, are both conserved in PRG-3 and PRG-5, in the case of His200 within the C2 domain of PRG-3, PRG-5 instead showed Glu195 (Tab. 1 and Fig. 4.22).

To understand if that residues, in PRG-5, have a role in the spine induction previously showed, I generated a series of proteins mutated at these single amino acids. The conservative amino acids Ser193 and Arg241 were exchanged with Trp193 and Glu241, respectively, while the non-conservative Glu195 was exchanged firstly with Gly195 and then with His195, to reproduce the same sequence of the PRG-3 C2 domain in PRG-5 (see also Tab. 1).

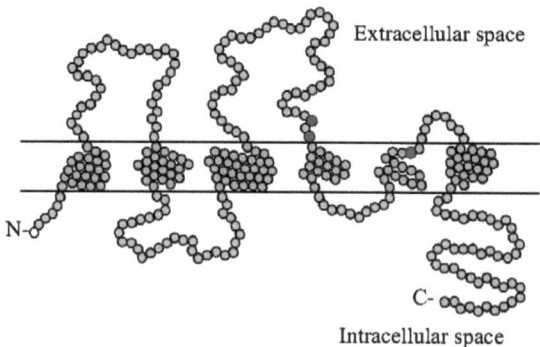

Fig. 4.22 Mouse PRG-5 structure. In red the three amino acids (Ser193, Glu195 and Arg241) exchanged. The double line represents the plasma membrane.

Western blot analyses of mutated PRG-5-Flag construct overexpressed in cultured HEK-293 cells confirmed expression of all mutants at the predicted molecular mass of 35 KDa (Fig. 4.23).

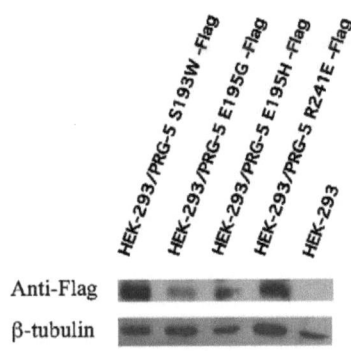

Fig. 4.23 Western blot analysis of mutant proteins detected with a flag-specific antibody.

Primary neurons were transfected with the same amount of PRG-5-flag wild type and PRG-5 Ser193Trp (PRG-5 S193W), -Glu195Gly (PRG-5 E195G), -Glu195His (PRG-5 E195H) and -Arg241Glu (PRG-5 R241E), fixed 24 hours later and investigated by confocal microscopy (Fig. 4.24). Primary neurons, overexpressing the PRG-5 protein with mutations at conservative amino acids S193W and R241E, as well non-conservative residue E195G, did not show any spines or membrane protrusions between DIV 2 and 4. Actin staining confirmed the absence of spines, filopodia or extracellular structures (Fig. 4.24 and 4.25).

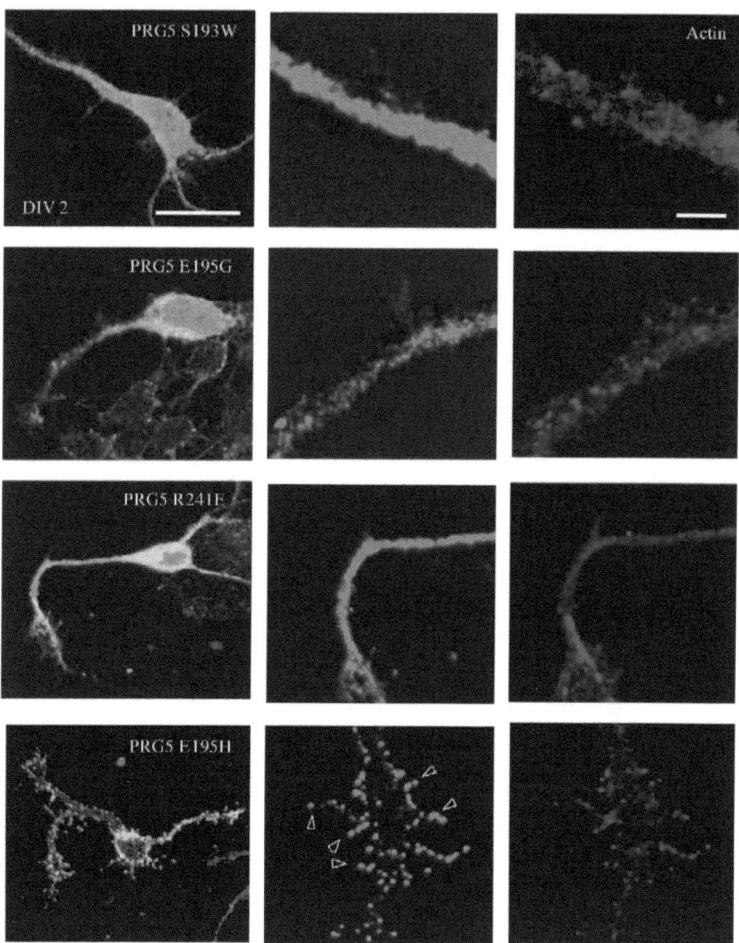

Fig. 4.24 Neurons at DIV 2 overexpressed the S193W, E195G, R241E and E195H PRG-5 mutants and investigated by confocal microscopy. Merge colors showed PRG-5-Flag mutans and actin staining. Scale bar represents 10 μm. Higher magnification of neuritis from a white box. PRG-5-Flag antibody (green, central panels) and actin staining (red, right panels). Scale bar represents 4μm.

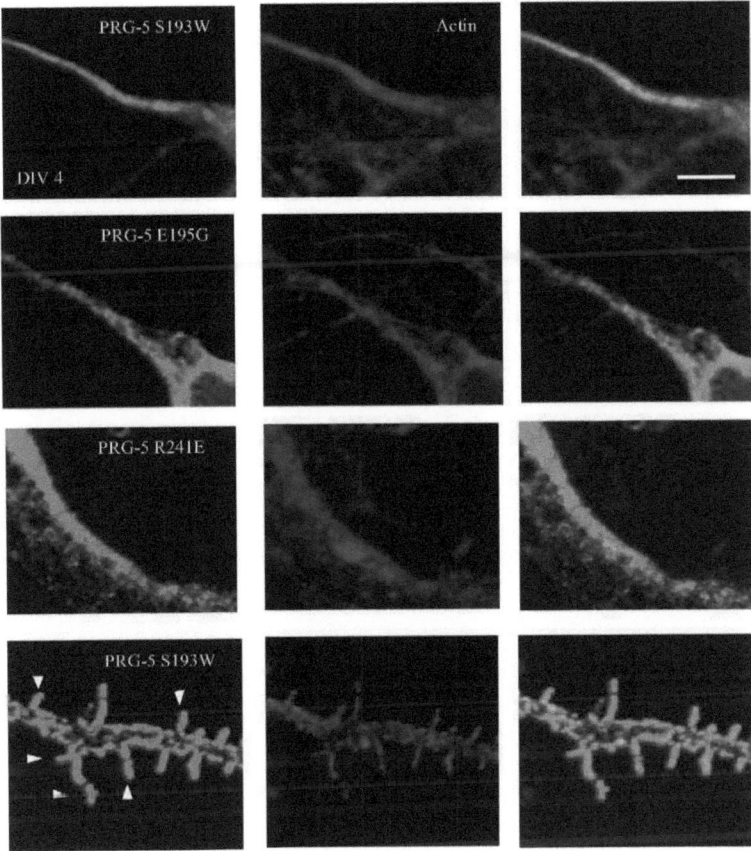

Fig. 4.25 Primary neurons at DIV 4, transfected with mutants S193W, E195G and R241E were unable to induce spines respect the mutant E195H. Mutants are stained in green while red showed actin staining. Scale bar represents 4µm.

In detail, primary neurons trasnfected with mutants S193W, E195G and R241E displayed a dispersed fluorescence signal in the intracellular region, partially in the soma and along the neurites (Fig. 4.24 and 4.25). This dispersed fluorescence could be explained as an intracellular localization of the mutant proteins. Exchange of the amino acids Ser193, Glu195 and Arg241 may interfere with a correct protein folding, blocking the transport of the self-mutated protein to the plasma membrane. Nevertheless, the mutant E195H, as the PRG-5-Flag wild type, induced spine formation in neurons at DIV 2 and 4, respectively (Fig. 4.24 and 4.25, bottom panels).

The overexpression of PRG-5-Flag E195H in primary neurons showed a considerable number of spine-like structures along the soma and the neuronal processes. Each spine was rich in actin structure and a clear fluorescent signal from the induced mutant protein was distinguishable. In addition, sequence alignment of different PRG-5 species (GeneBank) revealed that Glu195 was highly conserved within the C2 domain, while all PRG family members displayed Ser193 and Arg241 as highly conserved within the C2 and C3 domains, respectively (Tab. 1).

5. Discussion

The members of PRG family define a subclass of the LPP superfamily, characterized by six membrane-spanning domains containing three extracellular loops within highly conserved enzymatic active domains (6). LPP-1 is able to degrade LPA, controlling the level of lipids in the extracellular space (10, 29, 30, 50) and expression pattern of LPP-1 is been found in different tissues, included in the brain (27, 31, 63).

Here, I have investigated the upstream region of PRG-1, identified as proximal promoter and successively I characterized the last member of PRG family, plasticity-related gene-5 (PRG-5), able to induce spine-like structure in young primary neurons.

5.1 Proximal promoter of PRG-1

PRG-1 belongs to the family of plasticity related genes, which possesses homology to lipid phosphate phosphatases, and is part of the superfamily of lipid phosphate phosphatases/phosphotransferases (57). However, PRG-1 shows only a residual enzymatic activity for dephosphorylation of the superfamily specific substrates such as phosphatidate, lysophosphatidate or sphingosine-1-phosphate (10), which is due to amino acid exchanges in the catalytic motif that is shared by active PAP-2 family members (9). Principally, PRG-1 is involved in modulation of neuronal transmission via bioactive lipids, such as LPA acting via presynaptic LPA_2 receptors, which are controlled by PRG-1 from the postsynaptic side (specifically at the glutamatergic junction). According to recent data, this action is mediated by PRG-1 by depleting the pool of bioactive lipids that act on the presynaptic LPA_2 receptors (66).

The transcription of a eukaryotic gene is regulated by the combined action of multiple sequence-specific transcription factors, general transcription factors, histone modifiers, cofactors, and mediators that regulate transcription factor activity and chromatin structure. Our previous analysis of PRG-1 transcriptional regulation has shown that PRG-1 transcription is initiated at multiple transcription start sites. One sequence element 5'- at the translation start site (ATG) was detected. It mediates specific neuronal transcription in the human and rodent PRG-1 promoter (21a). Interesting to note is that PRG-1 is transcribed via a type "null" core promoter neither TATA box nor initiator element, and no CCAAT box. Multiple start sites are common for this promoter type (40). Computer aided analysis predicted also a 681 bp of a CpG island with 63.1% GC content 298 by upstream the first ATG of PRG-1. The existence of transcription start sites downstream of the first ATG of PRG-1 implies the existence of a second, alternative N-terminus. The shorter form of PRG-1 from second ATG needs to be confirmed, but resembles the homologous protein PRG-2 at the N-terminus (7), that is expressed prenatally and has a different expression pattern compared to PRG-1. Thus, the shorter form of PRG-1, lacking the N-terminal 49 amino acids, may substitutes the function of PRG-2 at postnatally stage. However, the function of N-terminual elongation remains to be clarified.

Recent study displayed the influence of Nex1, a homolog of the Drosophila proneural gene atonal belonging to the family of basic helix-loop-helix (bHLH) transcription factors, on PRG-1 expression. bHLH transcription factors bind to the consensus sequence CANNTG (E-Box), which has been identified in a variety of promoter and enhancer elements (35). Nex1 expression occurs at embryonic day 11.5 (55), and peaks in the first postnatal week parallel to dendritic arborisation and synaptogenesis of cortical neurons (2). Nex1 binds directly to at least on E-Box in the promoter region of PRG-1 and that it also mediates neurite outgrowth via PRG-1 expression in PC12 cells (72). Nex1 also regulates a molecular chaperone network through the expression of a set of heat shock proteins in the absence of stress, and connects neuronal differentiation with survival pathways (67). To determine the regulatory role of Nex1 in PRG-1 expression, I performed dual luciferase assays in primary cultured cells and subsequently analysed elements displaying activation regarding Nex1 regulation.

In details, we identified a 450 bp minimal promoter conferring enhanced specific transcription in neurons. Computer aided analysis for identification of transcription factor binding sites delivered several candidates on the 450 bp promoter mediating the neuronal stimulation of transcription. A Nex1 independent regulation of PRG-1which is mediated by a 450 bp promoter fragment, conferring enhanced specific transcription in neurons, was determined (21a). This regulation can be demonstrated under organotypic conditions, which underlines that this sequence is sufficient for the peculiar neuronal expression of PRG-1 in neurons. Moreover, a Sp1 binding sites, sequence-specific transcription factor that binds GC and GT boxes to activate a wide range of viral and cellular genes, were found inside the core promoter of PRG-1. Sp1 is important both in transcription initiation and activation, and it can be regulated by multiple mechanisms in a cell type-specific and promoter context specific manner (5). Sp1 has been linked to the maintenance of methylation-free CpG islands (34), and hypermethylation around Sp1 binding sites has been reported to reduce Sp1 binding, thereby decreasing transcription (75).

5.2 Plasticity-related gene 5 (PRG-5)

PRG-5 is an integral membrane protein containing 321 amino acids (aa) with both the N- and C-terminus present in the intracellular region that is representative of a class of five proteins with high homology to the lipid phosphate phosphatises (LPPs). Members of this class of proteins are characterized by highly conserved enzymatic active domains (6). LPP-1 is able to degrade LPA and control the level of lipids in the extracellular space (10, 29, 30, 50) and its expression pattern has been found in different tissues (31), including in the brain (27, 31, 63). Sequence alignment analysis revealed more than 95% similarity between human, rat and mouse PRG-5 and more than 80% homology with mPRG-3. Despite this, comparison of the relevant regions of PRG-3 and PRG-5 proteins with those of LPP family shows that crucial amino acids responsible for enzymatic ectophosphatase activity are exchanged (26, 61, 64, 74). The His171 and His223 within the C2 and C3 consensus domains of LPP-1, essential for enzymatic function, were replaced in PRG-5 with Glu195 and Asn247 respectively (Tab. 1). Consequently, PRG-5 should not have any ectophosphatase activity, as recently shown for PRG-3 (51, 58). However, overexpression of PRG-5 induced spine-like structures in young primary neurons.

Mainly expressed in the brain (11), I confirmed that PRG-5 mRNA was presented as early as embryonic day 14 (E14) and peaked at postnatal day 0 (P0), while the PRG-5 protein was detected only in plasma membrane fraction of neurons at DIV 2 and 4.

In neurons at DIV 2, 24 hours after transfection, PRG-5-Flag protein was localized to both the intracellular region and plasma membrane. Immunostaining using a marker for α-clathrin vesicle showed a partial co-localization between PRG-5-Flag and α-clathrin, which suggests that PRG-5 may is recruited from these vesicles in the cytosol region to be transported into the plasma membrane. In the other hand, the plasma membrane showed the presence of many spine-like structures, probably PRG-5-Flag induced.

Dendritic spines observed in fixed brain tissue shows various shapes, and are generally classified into three types: the thin type having a slender neck and a small head, the mushroom type having a short neck and a relatively large head, and the stubby type having no neck (Fig. 5.1). In living neurons, spine shapes easily interchange between the above three types. In other words, spine morphologies are snapshots of dynamic morphological changes. Therefore, not only the spine morphology but also its dynamic change should be elucidated to understand synaptic functions.

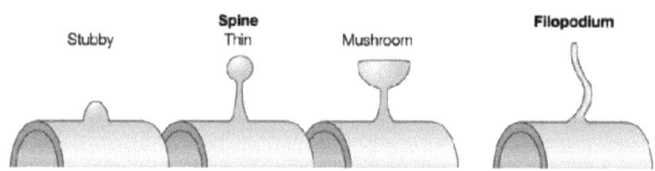

Fig 5.1 Schematic drawing of spine morphologies. Described in (73).

Hippocampal tissue prepared from P2-P7 animals and cultured 1-2 days in vitro generally show dendritic branches and lateral filopodia. When slices were cultured for 6-8 days in vitro, synaptogenesis and dendritic differentiation proceeded in vitro and spiny protrusions with complex shapes were clearly present (15). Dendrites of immature neurons, cultured for DIV 7 showed thin and long protrusions, clearly filopodia while neurons cultivated for DIV 21 were mature and covered with spines (54). On the other hand, dendrites of hippocampal neurons at DIV 10, transfected with EGFP, were covered with thin, long filopodia-like protrusions (44).

However, in our study, overexpression of PRG-5-Flag led to a large number of spines in neurons as young as DIV 2 and DIV 4. The spines were found with a mushroom and stubby shape, and the bulbous head contained PRG-5-Flag and polymerized actin. Filament of tubulin was detected only around the soma and along the neuritis, as expected. These types of membrane protrusions were not labelled when EGFP alone was expressed in these cells and suggest that a pathway for spine induction may exist that depend on PRG-5 expression. Stable spines could emerge on developing dendrites from pre-existing spiny structures or from dendrite shafts by *de novo* extension (15). In addition, newly formed spiny-like protrusions, developed over 48 hours, are morphologically indistinguishable from pre-existing spines, complicating the correct identification of the provenience of the mushroom spines already observed (76). Although our own observations are of primary neurons and not fixed tissue, it is interesting to note that neurons from E18 mice, cultivated at least 1 day in vitro, were able to produce mushroom and stubby spines after 24 h of overexpression of PRG-5, mainly along the neurites of the cell. This was still observable after 4 days in vitro, but not when the neurons were transfected with an empty EGFP as control.

Although the mechanism by which PRG-5 regulates the formation of spines requires further investigation, mutational studies provide insight into the possible mechanisms responsible. Unlike an earlier study, which analysed a truncated PRG-5 protein, in which the C-terminus, after deletion, did not allow for filopodia induction in N1E-115 cells (11), we investigated the "catalytic sites" of PRG-5 using single-point mutations, as previously reported for PRG-3. In details, PRG-3 overexpressed in COS7 and HeLa cells is able to increase the number of filopodia and the amino acids Ser193, His200 and Arg246, within the catalytic domains C1 – C3, were responsible for this function (58). However, despite the fact that PRG-5 displays high sequence homology with PRG-

3, it lacks critical residues present in the latter. In particular, instead of His200 within the C2 domain of PRG-3, PRG-5 instead shows Glu195 (Tab. 1).

Using direct mutagenesis, we substituted conservative Ser193 (C2 domain) and Arg241 (C3 domain) with Trp193 and Glu241 respectively and the non-conservative Glu195 (C2 domain), firstly in Gly195 and then in His195, in PRG-5 to replicate the C2 domain of PRG-3 (Tab. 1). The mutants S193W, R241E and E195G failed to induce spine formation in primary neurons at DIV 2 (Fig. 4.22) and 4 (Fig. 4.23). The predominant intracellular fluorescence distribution detected in the primary neurons could be the consequence of an incorrect folding of the mutated proteins, which may have hindered the transport of mutated PRG-5 to the plasma membrane. PRG-5 with the Glu195 mutated in His195, to reproduce the same C2 domain of PRG-3, was again able to induce spine formation in primary neurons.

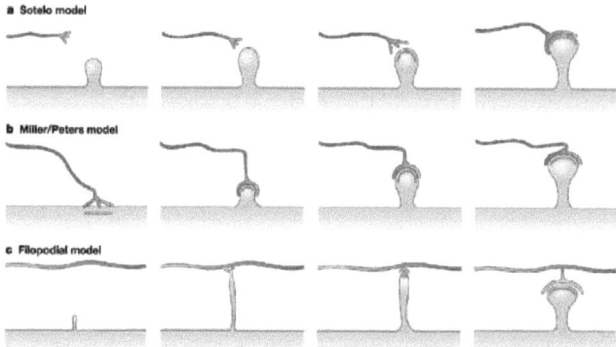

Fig. 5.2 This diagram illustrates the essential features of the three models of spinogenesis. In the Sotelo model (a), spines emerge independently of the axonal terminal. In the Miller/Peters model (b), the terminal actually induces the formation of the spine. Finally, in the filopodial model (c), a dendritic filopodium captures an axonal terminal and becomes a spine. (73).

Although the mechanisms involved in spine regulation remain unclear, three models for spinogenesis have been proposed, known as the Sotelo, Miller/Peters and Filopodia models (73) (Fig. 5.2). The Miller/Peters and Filopodia models suggest that a physical axon-dendrite interaction is required for the formation of spines (73), whereas the Sotelo model describes spines that emerge from dendrites independently of the axon. First, synapses are made on the dendritic shafts, and their flocculent material can recognize immature spines. Most of these spines are 'stubbies'. In the second stage, the presynaptic region of the axon shows a swelling as synaptic vesicles accumulate. In the third stage, many spines are thin or mushroom shaped, with a lollipop shape and a clear neck, and axonal terminals have well developed varicosities (Fig. 5.2 B and C).

This hypothesis is different from the Sotelo model (Fig. 5.2 A), which proposes that the terminal has a minor role in spinogenesis. It is possible that Purkinje cells and pyramidal neurons use completely different spinogenesis strategies, as they are very different cell types with seemingly different circuit functions (for example, excitatory versus inhibitory). At the same time, although there is consensus regarding the time course of events in pyramidal neurons, we feel that the available data on pyramidal spinogenesis are circumstantial and that the Miller/Peters hypothesis needs to be properly demonstrated (73).

PRG-5-Flag induces spine formation in neurites of immature primary neurons independently of an axon-dendrite interaction. Furthermore, our findings suggest that a necessary condition for PRG-5 to induce spines in primary neurons is a correct positioning of the protein at the plasma membrane. Given this, PRG-5, as a multispanning membrane protein, can induce a membrane deformation and consequently spine induction. A possible curvature of membrane was proposed for the I-BAR proteins (28). In the latter case, filopodia initiation could occur through membrane-deforming proteins that contain I-BAR (inverse Bin/Amphiphysin/Rvs) domains. Electrostatic interactions between the positively charged poles of I-BAR domains and the negatively charged $PI(4,5)P_2$ head groups induce clustering of $PI(4,5)P_2$ and generation of membrane curvature due to the convex geometry of the lipid-binding interface of the domain (28). Specifically in neurons, depletion of I-BAR on the protein IRSp53 affects dendritic spine morphogenesis (14). This is consistent with the observation that primary neurons transfected with PRG-5 mutated protein exhibit no spines, and that proteins were detected in the cytosol and partially in the membrane.

Moreover, a transient overexpression of PRG-5 in HEK-293 cells, resulted in a pronounced morphological change of the plasma membrane. PRG-5-Flag protein appeared packaged inside capsules that are all around the cells, as an incontrollable exocytosis (Fig. 5.3), may due to an extreme rearrangement of the plasma membrane.

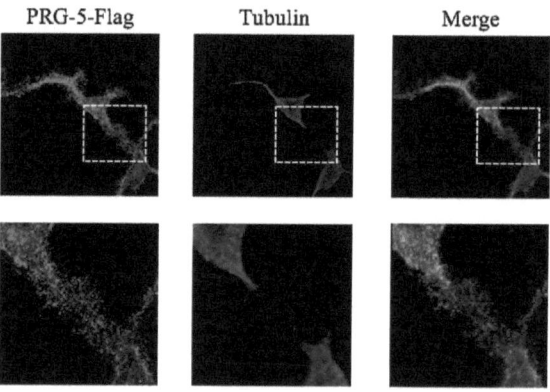

Fig. 5.3 HEK-293 cells transfected with pCDNA3-PRG-5-Flag (green) and analysed by confocal microscopy. Tubulin structure is visualized in red. Higher magnification of boxed area. Scale bar represents 10 and 4µm respectively.

In conclusion, we have found that mouse PRG-5, an integral transmembrane protein, is brain-specific and able to induce mushroom and stubby spines in immature primary neurons prior to the start of the physiological process of spinogenesis. Although we do not know the basis of this spine induction, our results suggest PRG-5 plays a critical role in spine induction, and that a basic prerequisite for this functioning is correct positioning at the plasma membrane.

6. Bibliography

1. **Andersen, P., and A. F. Soleng.** 1999. A thorny question: how does activity maintain dendritic spines? Nat Neurosci **2**:5-7.

2. **Bartholoma, A., and K. A. Nave.** 1994. NEX-1: a novel brain-specific helix-loop-helix protein with autoregulation and sustained expression in mature cortical neurons. Mech Dev **48**:217-28.

3. **Bonhoeffer, T., and R. Yuste.** 2002. Spine motility. Phenomenology, mechanisms, and function. Neuron **35**:1019-27.

4. **Bourne, J. N., and K. M. Harris.** 2008. Balancing structure and function at hippocampal dendritic spines. Annu Rev Neurosci **31**:47-67.

5. **Bouwman, P., and S. Philipsen.** 2002. Regulation of the activity of Sp1-related transcription factors. Mol Cell Endocrinol **195**:27-38.

6. **Brauer, A. U., and R. Nitsch.** 2008. Plasticity-related genes (PRGs/LRPs): a brain-specific class of lysophospholipid-modifying proteins. Biochim Biophys Acta **1781**:595-600.

7. **Brauer, A. U., N. E. Savaskan, H. Kuhn, S. Prehn, O. Ninnemann, and R. Nitsch.** 2003. A new phospholipid phosphatase, PRG-1, is involved in axon growth and regenerative sprouting. Nat Neurosci **6**:572-8.

8. **Bravin, M., L. Morando, A. Vercelli, F. Rossi, and P. Strata.** 1999. Control of spine formation by electrical activity in the adult rat cerebellum. Proc Natl Acad Sci U S A **96**:1704-9.

9. **Brindley, D. N.** 2004. Lipid phosphate phosphatases and related proteins: signalling functions in development, cell division, and cancer. J Cell Biochem **92**:900-12.

10. **Brindley, D. N., and D. W. Waggoner.** 1998. Mammalian lipid phosphate phosphohydrolases. J Biol Chem **273**:24281-4.

11. **Broggini, T., R. Nitsch, and N. E. Savaskan.** Plasticity-related gene 5 (PRG5) induces filopodia and neurite growth and impedes lysophosphatidic acid- and nogo-A-mediated axonal retraction. Mol Biol Cell **21**:521-37.

12. **Burke, T. W., and J. T. Kadonaga.** 1996. Drosophila TFIID binds to a conserved downstream basal promoter element that is present in many TATA-box-deficient promoters. Genes Dev **10**:711-24.

13. **Cesa, R., L. Morando, and P. Strata.** 2005. Purkinje cell spinogenesis during architectural rewiring in the mature cerebellum. Eur J Neurosci **22**:579-86.

14. **Choi, J., J. Ko, B. Racz, A. Burette, J. R. Lee, S. Kim, M. Na, H. W. Lee, K. Kim, R. J. Weinberg, and E. Kim.** 2005. Regulation of dendritic spine morphogenesis by insulin receptor substrate 53, a downstream effector of Rac1 and Cdc42 small GTPases. J Neurosci **25**:869-79.

15. **Dailey, M. E., and S. J. Smith.** 1996. The dynamics of dendritic structure in developing hippocampal slices. J Neurosci **16**:2983-94.

16. **Dotti, C. G., C. A. Sullivan, and G. A. Banker.** 1988. The establishment of polarity by hippocampal neurons in culture. J Neurosci **8**:1454-68.

17. **Dunaevsky, A., and C. A. Mason.** 2003. Spine motility: a means towards an end? Trends Neurosci **26**:155-60.

18. **Engert, F., and T. Bonhoeffer.** 1999. Dendritic spine changes associated with hippocampal long-term synaptic plasticity. Nature **399**:66-70.

19. **Fiala, J. C., M. Feinberg, V. Popov, and K. M. Harris.** 1998. Synaptogenesis via dendritic filopodia in developing hippocampal area CA1. J Neurosci **18**:8900-11.

20. **Fischer, M., S. Kaech, D. Knutti, and A. Matus.** 1998. Rapid actin-based plasticity in dendritic spines. Neuron **20**:847-54.

21. **Fourcade, O., M. F. Simon, C. Viode, N. Rugani, F. Leballe, A. Ragab, B. Fournie, L. Sarda, and H. Chap.** 1995. Secretory phospholipase A2 generates the novel lipid mediator lysophosphatidic acid in membrane microvesicles shed from activated cells. Cell **80**:919-27.

21a. **Geist B., Vorwerk B., Coiro P., Ninnemann O., and Nitsch R.** 2012. PRG-1 transcriptional regulation independent from Nex1/Math2-mediated activation. Cell Moll Life Sci **69**:651-61.

22. **Grutzendler, J., N. Kasthuri, and W. B. Gan.** 2002. Long-term dendritic spine stability in the adult cortex. Nature **420**:812-6.

23. **Halpain, S.** 2000. Actin and the agile spine: how and why do dendritic spines dance? Trends Neurosci **23**:141-6.

24. **Hampsey, M.** 1998. Molecular genetics of the RNA polymerase II general transcriptional machinery. Microbiol Mol Biol Rev **62**:465-503.

25. **Harris, K. M.** 1999. Structure, development, and plasticity of dendritic spines. Curr Opin Neurobiol **9**:343-8.

26. **Hemrika, W., R. Renirie, H. L. Dekker, P. Barnett, and R. Wever.** 1997. From phosphatases to vanadium peroxidases: a similar architecture of the active site. Proc Natl Acad Sci U S A **94**:2145-9.

27. **Hooks, S. B., S. P. Ragan, and K. R. Lynch.** 1998. Identification of a novel human phosphatidic acid phosphatase type 2 isoform. FEBS Lett **427**:188-92.

28. **Hotulainen, P., and C. C. Hoogenraad.** Actin in dendritic spines: connecting dynamics to function. J Cell Biol **189**:619-29.

29. **Ishikawa, T., M. Kai, I. Wada, and H. Kanoh.** 2000. Cell surface activities of the human type 2b phosphatidic acid phosphatase. J Biochem **127**:645-51.

30. **Jasinska, R., Q. X. Zhang, C. Pilquil, I. Singh, J. Xu, J. Dewald, D. A. Dillon, L. G. Berthiaume, G. M. Carman, D. W. Waggoner, and D. N. Brindley.** 1999. Lipid phosphate phosphohydrolase-1 degrades exogenous glycerolipid and sphingolipid phosphate esters. Biochem J **340 (Pt 3)**:677-86.

31. **Kai, M., I. Wada, S. Imai, F. Sakane, and H. Kanoh.** 1997. Cloning and characterization of two human isozymes of Mg2+-independent phosphatidic acid phosphatase. J Biol Chem **272**:24572-8.

32. **Kim, Y., J. Y. Sung, I. Ceglia, K. W. Lee, J. H. Ahn, J. M. Halford, A. M. Kim, S. P. Kwak, J. B. Park, S. Ho Ryu, A. Schenck, B. Bardoni, J. D. Scott, A. C. Nairn, and P. Greengard.** 2006. Phosphorylation of WAVE1 regulates actin polymerization and dendritic spine morphology. Nature **442**:814-7.

33. **Lohmann, C., and T. Bonhoeffer.** 2008. A role for local calcium signaling in rapid synaptic partner selection by dendritic filopodia. Neuron **59**:253-60.

34. **Macleod, D., J. Charlton, J. Mullins, and A. P. Bird.** 1994. Sp1 sites in the mouse aprt gene promoter are required to prevent methylation of the CpG island. Genes Dev **8**:2282-92.

35. **Massari, M. E., and C. Murre.** 2000. Helix-loop-helix proteins: regulators of transcription in eucaryotic organisms. Mol Cell Biol **20**:429-40.

36. **Matus, A.** 1999. Postsynaptic actin and neuronal plasticity. Curr Opin Neurobiol **9**:561-5.

37. **Moolenaar, W. H.** 1995. Lysophosphatidic acid signalling. Curr Opin Cell Biol **7**:203-10.

38. **Nakayama, A. Y., M. B. Harms, and L. Luo.** 2000. Small GTPases Rac and Rho in the maintenance of dendritic spines and branches in hippocampal pyramidal neurons. J Neurosci **20**:5329-38.

39. **Nanjundan, M., and F. Possmayer.** 2001. Pulmonary lipid phosphate phosphohydrolase in plasma membrane signalling platforms. Biochem J **358**:637-46.

40. **Novina, C. D., and A. L. Roy.** 1996. Core promoters and transcriptional control. Trends Genet **12**:351-5.

41. **Orphanides, G., T. Lagrange, and D. Reinberg.** 1996. The general transcription factors of RNA polymerase II. Genes Dev **10**:2657-83.

42. **Parnass, Z., A. Tashiro, and R. Yuste.** 2000. Analysis of spine morphological plasticity in developing hippocampal pyramidal neurons. Hippocampus **10**:561-8.

43. **Peeva, G. P., S. K. Angelova, O. Guntinas-Lichius, M. Streppel, A. Irintchev, U. Schutz, A. Popratiloff, N. E. Savaskan, A. U. Brauer, A. Alvanou, R. Nitsch, and D. N. Angelov.** 2006. Improved outcome of facial nerve repair in rats is associated with enhanced regenerative response of motoneurons and augmented neocortical plasticity. Eur J Neurosci **24**:2152-62.

44. **Penzes, P., A. Beeser, J. Chernoff, M. R. Schiller, B. A. Eipper, R. E. Mains, and R. L. Huganir.** 2003. Rapid induction of dendritic spine morphogenesis by trans-synaptic ephrinB-EphB receptor activation of the Rho-GEF kalirin. Neuron **37**:263-74.

45. **Penzes, P., R. C. Johnson, R. Sattler, X. Zhang, R. L. Huganir, V. Kambampati, R. E. Mains, and B. A. Eipper.** 2001. The neuronal Rho-GEF Kalirin-7 interacts with PDZ domain-containing proteins and regulates dendritic morphogenesis. Neuron **29**:229-42.

46. **Pilquil, C., J. Dewald, A. Cherney, I. Gorshkova, G. Tigyi, D. English, V. Natarajan, and D. N. Brindley.** 2006. Lipid phosphate phosphatase-1 regulates lysophosphatidate-induced fibroblast migration by controlling phospholipase D2-dependent phosphatidate generation. J Biol Chem **281**:38418-29.

47. **Pilquil, C., Z. C. Ling, I. Singh, K. Buri, Q. X. Zhang, and D. N. Brindley.** 2001. Co-ordinate regulation of growth factor receptors and lipid phosphate phosphatase-1 controls cell activation by exogenous lysophosphatidate. Biochem Soc Trans **29**:825-30.

48. **Portera-Cailliau, C., D. T. Pan, and R. Yuste.** 2003. Activity-regulated dynamic behavior of early dendritic protrusions: evidence for different types of dendritic filopodia. J Neurosci **23**:7129-42.

49. **Pyne, S., K. C. Kong, and P. I. Darroch.** 2004. Lysophosphatidic acid and sphingosine 1-phosphate biology: the role of lipid phosphate phosphatases. Semin Cell Dev Biol **15**:491-501.

50. **Roberts, R. Z., and A. J. Morris.** 2000. Role of phosphatidic acid phosphatase 2a in uptake of extracellular lipid phosphate mediators. Biochim Biophys Acta **1487**:33-49.

51. **Savaskan, N. E., A. U. Brauer, and R. Nitsch.** 2004. Molecular cloning and expression regulation of PRG-3, a new member of the plasticity-related gene family. Eur J Neurosci **19**:212-20.

52. **Savaskan, N. E., and R. Nitsch.** 2001. Molecules involved in reactive sprouting in the hippocampus. Rev Neurosci **12**:195-215.

53. **Sciorra, V. A., and A. J. Morris.** 1999. Sequential actions of phospholipase D and phosphatidic acid phosphohydrolase 2b generate diglyceride in mammalian cells. Mol Biol Cell **10**:3863-76.

54. **Sekino, Y., N. Kojima, and T. Shirao.** 2007. Role of actin cytoskeleton in dendritic spine morphogenesis. Neurochem Int **51**:92-104.

55. **Shimizu, C., C. Akazawa, S. Nakanishi, and R. Kageyama.** 1995. MATH-2, a mammalian helix-loop-helix factor structurally related to the product of Drosophila proneural gene atonal, is specifically expressed in the nervous system. Eur J Biochem **229**:239-48.

56. **Shoukimas, G. M., and J. W. Hinds.** 1978. The development of the cerebral cortex in the embryonic mouse: an electron microscopic serial section analysis. J Comp Neurol **179**:795-830.

57. **Sigal, Y. J., M. I. McDermott, and A. J. Morris.** 2005. Integral membrane lipid phosphatases/phosphotransferases: common structure and diverse functions. Biochem J **387**:281-93.

58. **Sigal, Y. J., O. A. Quintero, R. E. Cheney, and A. J. Morris.** 2007. Cdc42 and ARP2/3-independent regulation of filopodia by an integral membrane lipid-phosphatase-related protein. J Cell Sci **120**:340-52.

59. **Skutella, T., and R. Nitsch.** 2001. New molecules for hippocampal development. Trends Neurosci **24**:107-13.

60. **Smale, S. T., and J. T. Kadonaga.** 2003. The RNA polymerase II core promoter. Annu Rev Biochem **72**:449-79.

61. **Stukey, J., and G. M. Carman.** 1997. Identification of a novel phosphatase sequence motif. Protein Sci **6**:469-72.

62. **Sun, L., S. Gu, Y. Sun, D. Zheng, Q. Wu, X. Li, J. Dai, C. Ji, Y. Xie, and Y. Mao.** 2005. Cloning and characterization of a novel human phosphatidic acid phosphatase type 2, PAP2d, with two different transcripts PAP2d_v1 and PAP2d_v2. Mol Cell Biochem **272**:91-6.

63. **Suzuki, R., H. Sakagami, Y. Owada, Y. Handa, and H. Kondo.** 1999. Localization of mRNA for Dri 42, subtype 2b of phosphatidic acid phosphatase, in the rat brain during development. Brain Res Mol Brain Res **66**:195-9.

64. **Toke, D. A., M. L. McClintick, and G. M. Carman.** 1999. Mutagenesis of the phosphatase sequence motif in diacylglycerol pyrophosphate phosphatase from Saccharomyces cerevisiae. Biochemistry **38**:14606-13.

65. **Trachtenberg, J. T., B. E. Chen, G. W. Knott, G. Feng, J. R. Sanes, E. Welker, and K. Svoboda.** 2002. Long-term in vivo imaging of experience-dependent synaptic plasticity in adult cortex. Nature **420**:788-94.

66. **Trimbuch, T., P. Beed, J. Vogt, S. Schuchmann, N. Maier, M. Kintscher, J. Breustedt, M. Schuelke, N. Streu, O. Kieselmann, I. Brunk, G. Laube, U. Strauss, A. Battefeld, H. Wende, C. Birchmeier, S. Wiese, M. Sendtner, H. Kawabe, M. Kishimoto-Suga, N. Brose, J. Baumgart, B. Geist, J. Aoki, N. E. Savaskan, A. U. Brauer, J. Chun, O. Ninnemann, D. Schmitz, and R. Nitsch.** 2009. Synaptic PRG-1 modulates excitatory transmission via lipid phosphate-mediated signaling. Cell **138**:1222-35.

67. **Uittenbogaard, M., K. K. Baxter, and A. Chiaramello.** NeuroD6 genomic signature bridging neuronal differentiation to survival via the molecular chaperone network. J Neurosci Res **88**:33-54.

68. **Umezu-Goto, M., Y. Kishi, A. Taira, K. Hama, N. Dohmae, K. Takio, T. Yamori, G. B. Mills, K. Inoue, J. Aoki, and H. Arai.** 2002. Autotaxin has lysophospholipase D activity leading to tumor cell growth and motility by lysophosphatidic acid production. J Cell Biol **158**:227-33.

69. **Wong, W. T., and R. O. Wong.** 2000. Rapid dendritic movements during synapse formation and rearrangement. Curr Opin Neurobiol **10**:118-24.

70. **Woychik, N. A., and M. Hampsey.** 2002. The RNA polymerase II machinery: structure illuminates function. Cell **108**:453-63.

71. **Xie, Y., J. P. Vessey, A. Konecna, R. Dahm, P. Macchi, and M. A. Kiebler.** 2007. The GTP-binding protein Septin 7 is critical for dendrite branching and dendritic-spine morphology. Curr Biol **17**:1746-51.

72. **Yamada, M., Y. Shida, K. Takahashi, T. Tanioka, Y. Nakano, and T. Tobe.** 2008. Prg1 is regulated by the basic helix-loop-helix transcription factor Math2. J Neurochem **106**:2375-84.

73. **Yuste, R., and T. Bonhoeffer.** 2004. Genesis of dendritic spines: insights from ultrastructural and imaging studies. Nat Rev Neurosci **5**:24-34.

74. **Zhang, Q. X., C. S. Pilquil, J. Dewald, L. G. Berthiaume, and D. N. Brindley.** 2000. Identification of structurally important domains of lipid phosphate phosphatase-1: implications for its sites of action. Biochem J **345 Pt 2**:181-4.

75. **Zhu, W. G., K. Srinivasan, Z. Dai, W. Duan, L. J. Druhan, H. Ding, L. Yee, M. A. Villalona-Calero, C. Plass, and G. A. Otterson.** 2003. Methylation of adjacent CpG sites affects Sp1/Sp3 binding and activity in the p21(Cip1) promoter. Mol Cell Biol **23**:4056-65.

76. **Zuo, Y., A. Lin, P. Chang, and W. B. Gan.** 2005. Development of long-term dendritic spine stability in diverse regions of cerebral cortex. Neuron **46**:181-9.

i want morebooks!

Buy your books fast and straightforward online - at one of world's fastest growing online book stores! Environmentally sound due to Print-on-Demand technologies.

Buy your books online at
www.get-morebooks.com

Kaufen Sie Ihre Bücher schnell und unkompliziert online – auf einer der am schnellsten wachsenden Buchhandelsplattformen weltweit! Dank Print-On-Demand umwelt- und ressourcenschonend produziert.

Bücher schneller online kaufen
www.morebooks.de

VDM Verlagsservicegesellschaft mbH
Heinrich-Böcking-Str. 6-8　　Telefon: +49 681 3720 174　　info@vdm-vsg.de
D - 66121 Saarbrücken　　　Telefax: +49 681 3720 1749　　www.vdm-vsg.de

Printed by Books on Demand GmbH, Norderstedt / Germany